Biocidal

Confronting the Poisonous Legacy of PCBs

TED DRACOS

Beacon Press
Boston

Beacon Press
25 Beacon Street
Boston, Massachusetts 02108-2892
www.beacon.org

Beacon Press books
are published under the auspices of
the Unitarian Universalist Association of Congregations.

13 12 11 10   8 7 6 5 4 3 2 1

This book is printed on acid-free paper that meets the uncoated paper
ANSI/NISO specifications for permanence as revised in 1992.

Text design and composition by Wilsted & Taylor Publishing Services

Library of Congress Cataloging-in-Publication Data
Dracos, Ted.
Biocidal : confronting the poisonous legacy of PCBs / Ted Dracos.
p. ; cm.
Includes bibliographical references and index.
ISBN 978-0-8070-0612-2 (hardcover : alk. paper) 1. Polychlorinated biphenyls—
Toxicology. 2. Polychlorinated biphenyls—Environmental effect. I. Title.
[DNLM: 1. Polychlorinated Biphenyls—history—United States. 2. Polychlorinated
Biphenyls—poisoning—United States. 3. Environmental Pollutants—history—United
States. 4. Environmental Pollutants—poisoning—United States. 5. History, 20th
Century—United States. WA 11 AA1 D757b 2010]
RA1242.P7D73 2010
363.7—dc22   2010013432

Excerpt from *We Are Three: New Rumi Poems* (Maypop Books, 1988),
translated by Coleman Barks, reprinted by permission.

Excerpt from Mary O'Brien's presentation to the Society for Risk Analysis,
"Beyond Democratization of Risk Assessment: An Alternative to
Risk Assessment" (2001), reprinted by permission.

For Lilah

Outside, the freezing desert night.
This other night inside grows warm, kindling.
Let the landscape be covered with thorny crust.
We have a soft garden in here.
The continents blasted,
cities and little towns, everything
become a scorched, blackened ball.

The news we hear is full of grief for that future,
but the real news inside here
is there's no news at all.

JELALUDDIN RUMI

Contents

CHAPTER I

The Man Who Poisoned the Planet

Gatsby believed in the green light, the orgastic future that year by year
recedes before us. It eluded us then, but that's no matter—tomorrow we
will run faster, stretch out our arms farther. . . . And one fine morning—

F. SCOTT FITZGERALD, *The Great Gatsby*, 1925

There is no record of whether there was a wager involved or if it was just a dare. Whichever it was, when Theodore Swann returned home to Anniston, Alabama, after the 1920 running of the Kentucky Derby, the thirty-three-year-old industrialist had already drawn up an insane challenge for his young chemists. He gave them sixty days to manufacture two tank cars full of a substance they would dub the "magic fluid"—a chemical compound that Swann believed would change not only America, but eventually the world.

That first year of the Roaring Twenties led the way, propagating mythical heroes and creating vast social transformations in America. Radio station KDKA of Pittsburgh made the first commercial media broadcasts, even though its audience was mostly wearing headphones plugged into primitive radios that used a chunk of rock crystal for tuning. Margaret Gorman was crowned the first Miss America in Atlantic City over the protests of religious groups scandalized by the bare arms and calves of the contestants. Despite the success of the racy Miss America pageant, the country's politically powerful moral crusaders were in full cry and enjoyed their greatest victory ever in the first year of that wild decade: the Eighteenth Amendment went into effect and the sale of alcohol in the United States became illegal.

And in the summer of 1920, without ceremony, Theodore Swann's chemists produced the first commercial quantities of a substance that catalyzed the most revolutionary cultural event of the entire twentieth century—the launching of the fabled American lifestyle, along with the unforeseen poisoning of the planet.

Sixty days. Perhaps Swann's chemists thought he'd had one too many bootlegged mint juleps when he issued this challenge. All the chemists really knew was that the "magic fluid" had never been mass produced. Eastman Kodak was selling some in small amounts, but at the exorbitant price of forty dollars per pound. Swann's chemists weren't even sure what raw materials were needed to create the substance, let alone the proportions or temperatures required for mixing and producing it.

All Swann knew was that massive amounts of the material were needed by power companies that were electrifying the South along with the rest of America. From his executive-level contacts he heard they were desperate for a stable and cheap heat-transfer material that could keep their huge transformers cool and thus prevent explosions. As far as the power companies were concerned, without such a substance, there would be no electrical grid to power the United States.

Theodore Swann, as geniuses often do, attracted talented and loyal men to work for him, both scientists and blue-collar workers. The young entrepreneur also had a knack for pushing the motivational buttons of these men, by using a slick combination of psychology and financial incentives. First, he would give them some over-the-top production goal and tell them there was no way they could ever do it. Then he'd pay them inflated wages with "pass-off" chits—a system Swann used as soon as he began operating his own companies. These chits were essentially a forerunner of stock options, which employees were entitled to cash in for money or stock at a later date.

By combining his off-the-wall challenges with "pass-offs," Swann got his elite production crews to work for seventy-two-hour stretches.

The schedule was so brutal that his men would sleep in their wheelbarrows with legs stretched out on the handles. Swann's sleeping system for the crews was straightforward: you could sleep anytime you wanted, but you had to sleep in the open where you could be seen by everybody. That way the men could observe for themselves if the load was being equally shared.

Sixty days! This was not only a preposterous deadline for Swann's production gangs, but it sounded equally impossible to his young chemists. The specifications their chief outlined seemed pure fantasy. First, the process had to produce a compound that wouldn't clog up Swann's large and expensive new electric furnaces. He had also required that the "magic fluid" be made in large quantities and, at the same time, safe to handle. And it could have no waste byproducts—no emissions—that might rile up the good citizens of Anniston, Alabama.

Always-creative German chemists had stumbled on the "magic fluid" in the 1880s. They had found that the compound, a derivative of coal tar, indeed had magical qualities that would be highly prized by the world's burgeoning heavy industries. The mysterious material resisted breakdown from heat to a degree never before seen *and* was also the most efficient non-conductor of electricity that scientists had ever tested. But the problem was that no one could manufacture it at a reasonable cost. Eastman Kodak had developed a process for making the compound, but it could hardly be called "manufacturing," since they were selling it at such a high price. And Swann wanted railroad car-full batches of the stuff!

Even though industrial chemistry in the 1920s was a rough-and-ready business, a kind of industrial cowboying, Swann's chemists were gung ho. Safety precautions were somewhere between primitive and nonexistent—in fact, safety concerns were thought to be for sissies. Working long hours, Swann's men had to figuratively ride bareback, mixing batches of chemicals, heating them, cooling them, and then often analyzing the results on the fly.

The most aggressive of Swann's chemical wranglers thought they might be able to make the "magic fluid" by heating a vapor made from coal and then quickly cooling it in order to capture the precipitate. They scrounged equipment and scrap parts here and there, quickly building an experimental contraption—basically a large vat with a screened pipe that held a heating element inside.

Despite some minor explosions, a tiny bit of pure product was obtained by the anxious young chemists. But there was a hitch: The heating element had become badly clogged. Every time Swann's chemists tried to make the "magic fluid" in quantity, they triggered little explosions and then the guts of the cooking vat would be covered with a mass of lava-like amalgam. The dog-tired chemists had to face the truth: making two tank cars full of "magic fluid" this way seemed impossible.

In near desperation, one of the chemists came up with a bold idea involving the main ingredient, benzene. He believed they were not heating the benzene to a high enough temperature and then not cooling it fast enough. If they could superheat the substance and then supercool it, then they might have a shot at capturing a large enough volume of "magic fluid" to meet Swann's deadline.

But heating volatile liquids like benzene into vapors and then rapidly cooling them after introducing them to other chemicals—well, that was a formula for a *big* explosion. Swann's chemists knew this, of course, but with their deadline looming they were becoming frantic. Wrestling with serious doubts, the experimenters built a funky, blast-proof observation area out of sandbags, where they would hide when the test was conducted.

Soon all was ready. Heat was applied to the appropriate elements of the jury-rigged device. As the benzene began to boil and fumes started exiting from the big vat into the pipe precipitator, the men retreated to their sandbagged redoubt to await results. It seemed to be taking too long, though. Just as the team leader was about to peek over the safety barrier, there was an enormous explosion. Glass and lead shrapnel flew everywhere, followed instantaneously by a strong blast wave.

The flash fire subsided, leaving the uninjured but stunned chemists to ponder what had gone wrong. After sifting through the wreckage, they quickly realized that their heating techniques had left much to be desired. But in reviewing their procedures, they agreed that the principle of raising benzene to a high temperature and then rapidly cooling it was probably the secret to making the substance in quantity. Within a few days they had refined their methods to the point where they could actually begin manufacturing the "magic fluid" in magnum amounts.

Theodore Swann and his talented young scientists had made history. Commercially viable polychlorinated biphenyl, PCB, was created. Just at the right time, they came up with a means of mass-producing a substance, perhaps *the* critical substance, for an American and global industry that would change the quality of human existence. Of course they had no idea that they'd also created a Frankenstein chemical.

The "magic fluid," in one or more of its 209 varieties, would course through the blood and reside in the tissues of every human being ever tested for it. PCBs would invade every corner of the world, contaminating even the most remote landscapes. Entire species would quietly succumb to infinitesimal amounts of Swann's hardy, lethal, and mysterious molecules and their byproducts.

Not only that, the chemical legacy spawned by the Kentucky Derby wager would become part of Earth's environment for millennia. But unlike their toxic cousin DDT, the widely applied pesticide, PCBs were never released intentionally. They were to become the most deadly single toxin inadvertently released into the biosphere.

The man who was responsible for the unknowing chemical contamination of the globe was born on September 5, 1886, in Dandridge, Tennessee, to John S. Swann and his wife, Sonora, a woman reputedly as lovely as her melodious name. Never rich but relatively prosperous, John Swann descended from a long line of southern farmers, small businessmen, and sometime soldiers. He maintained the sacred tradition by fighting for the Confederacy in the Civil War as an infantryman. Although many of the Tennessee Swanns fought for the Northern side, John remained a defender of the old way of life that his father enjoyed as the owner of two slaves.

Young Theodore would listen intently to his father's war stories—the bloody, terrible battles at Shiloh, Chickamauga, and Missionary Ridge; the time toward the end of the war that General Robert E. Lee passed up the line of Swann's cheering regiment riding a fine charger and wearing an immaculate uniform (even though he'd slept hidden in a haystack the previous night to avoid marauding Union cavalry). As a result, the young Theodore was drenched in the lore and traditions of the South: courage, loyalty, gallantry, and an antipathy toward segments of northern culture that would play a part in his rise—and fall—as an industrialist.

After a lackluster, mostly religion-dominated high school and college education, Swann showed immediate skill at business, as well as leadership, by forming a small company to perform electrical contracting work for Westinghouse Electric in Bluefield, Virginia. While juggling contracting duties, Swann located a prospective customer, West Virginia Light and Traction Company, for a Westinghouse locomotive. Selling a huge steam locomotive then would be like selling a 747 jumbo jet today. But the young businessman not only made and closed the deal; he sold the West Virginia power company *two* locomotives. And the Swann legend was born.

In 1912, Swann moved to Charleston, West Virginia, and became the commercial manager for the newly formed Virginia Power Company, acting as their chief salesman of wholesale electrical power. Within months he had sold so many contracts that the company had to double its generating capacity to meet demand. Not surprisingly, Swann caught the eye of the president of Alabama Power, James Mitchell, one of the most influential business leaders in the state. Mitchell asked Swann to join his company in Birmingham and organize and run their corporate sales department. Swann accepted the offer.

By the outbreak of World War I, Swann had been appointed vice president of two companies, manufacturing steel and ordnance, respectively, in Anniston, Alabama. With no prior experience, he supervised an entire production run of heavy artillery shells, all the while commuting the seventy-five miles to Anniston from Birmingham on a two-lane mostly dirt road.

With the ordnance works operating at full capacity, the army asked Swann if he might make military-grade phosphorus for the government. Phosphorus burns and produces smoke as soon as it comes in contact with air, and the War Department wanted to obtain a large supply for smoke screens to cover troops that had to advance in open fields on the Western Front in Europe. Swann agreed, but the war ended before he could manufacture the material in quantity.

Even so, the phosphorus deal was to catapult him into the industrialist elite. With the War Department having approached him to do business, Swann decided he needed to set up his own, personally held chemical company—and for that, he needed working capital. Regrettably, his newly formed company, the Southern Manganese Corporation, didn't have much business making phosphorus, so Swann had to come up with a ploy to get more customers.

He decided to dupe a group of Eastern bankers with an industrial-strength variation on one of man's oldest scams, a sort of shell game. He would script a scene to convince the New Yorkers that Southern Manganese was a going concern that had more business than it knew what to do with. His problem, besides a lack of customers, was that his ferrophosphorus furnaces, even going full blast, never produced dramatic amounts of material such as the rivers of molten ore and meteoric sparks that a steel mill might theatrically disgorge for onlookers.

Accordingly, Swann would have to exaggerate his operations a bit. The day before the bankers arrived he would have all his boilers plugged. Then on the day the investors toured the facility, he would unplug them and boilers would gush forth their industrial chemical produce in prodigious, if artificially elevated, quantities. Or at least Swann hoped they would.

But the preparations for this were heavy with danger. The plugged, heated boilers could explode, not only destroying his new factory, but killing or maiming many of his best workers. Even so, Swann presented the plan to his select production crew, telling them that his scheme was critical, but completely voluntary. He also added a twist: they would have to go from boiler room to boiler room, since Swann didn't have enough workers to operate each boiler separately. He just had to hope that the bankers wouldn't notice that they were seeing the same men at every stop.

Swann's troopers agreed and all was made ready. He picked up the New Yorkers in Birmingham in a chauffeured Packard sedan well stocked with aged bourbon and fine scotch for the drive to Anniston. After settling them in a private club where he had arranged posh accommodations, he drove them to his executive office for more drinks before the plant tour.

When the excursion finally began, the operation was humming with convincing activity and, luckily for Swann, veritable rivers of product. By the eighth boiler, the bankers had signed on and Swann had the working capital he needed for his grandiose visions—visions that would be helped by a stroke of bizarre luck as much as by the northern dollars.

After getting his capital infusion, Swann started mass production of ferrophosphorus, an additive needed by the ravenous American steel industry. Production went well and markets were strong. But the manufacturing process became unendurable for the people of Anniston. Swann's furnaces were belching toxic dust and smoke around the clock, and the city was being eaten up by pollution as lawn furniture corroded, house paint blistered, and automobile paint jobs pitted overnight.

After meeting with irate townsfolk, Swann agreed to install "scrubbers"—at the time, state-of-the-art pollution control devices—on his smoke stacks. After the first day of operation, one of Swann's chemists took the clear, syrupy liquid byproduct from the scrubbers back to the laboratory for analysis. What he found bordered on the chemically amazing. Swann's pollution control scrubbers had serendipitously made pure, concentrated phosphoric acid.

But such a breakthrough, while exciting scientifically, was meaningless from a commercial perspective unless a market for their discovery could be found. Again Swann's luck held. There wasn't just *a* market, there were *many* markets—and they were potentially huge. One of the first to recognize the value of Swann's pure phosphoric acid was the Coca-Cola Company, already on its way to becoming an international business giant and located just a hundred miles away in Atlanta. They couldn't get enough of it for their top-secret soft-drink formula.

There were other large markets to till as well. Phosphorus was going to become a key ingredient in the most widely produced chemical product of modern times, agricultural fertilizer. Swann's lucky discovery would also be used as the central ingredient in home detergents—later to be voluntarily removed by manufacturers in the seventies after devastating aquatic ecosystems in one of America's first and worst environmental disasters.

But that was far in the future. By 1929, with his plants gushing out huge volumes of phosphorus as well as PCBs for electric companies that only clamored for more, at just about any price he wanted to set, Swann was a very rich man, and one headed straight for self-induced ruin.

While the Roaring Twenties, with its fabled Gatsby-esque excesses, certainly may have offered a cultural setting that contributed to Swann's decline, the era can hardly be blamed. Whatever the cause, the fall of the genius industrialist would be steep, fraught with folly, and approaching—if not breaching—criminality.

Whether it was his intention to build the most superlative residence Alabama had ever seen isn't known. Nonetheless, Swann's five-story castle would be that and more, unsurpassed not only in opulence, but in pretension as well. During his frequent European travels in the mid-twenties, Swann had fallen in love with a gloomy Norman castle, Hedingham Keep, northeast of London. With the profits from PCBs and phosphorus bursting his bank accounts, he hired a team of prominent architects to supervise reproduction of the aged monstrosity, right down to the original flaws by its medieval builders.

While the exterior was true to its heritage, the interior of Swann's Castle, as it was known, was a mishmash—starring an exact replica of Hedingham Keep's cavernous dining room, called Norman Hall (allegedly built by the son-in-law of William the Conqueror). Swann spared no expense for its fittings and fixtures, either. He brought back loads of crude medieval English furnishings. When authentic materials couldn't be found, he had them made. In the castle's dining hall hung "genuine" reproductions of the coats of arms of the knights of William's army. They overlooked a giant wooden banquet table, weighing close to half a ton, at the head of which was an ornately carved, high-backed throne chair befitting an industrial monarch.

With the Crash of 1929 and the country clenched by the Great Depression, Swann remained buoyant and refused to restrict his profligate ways. He not only completed the work on his castle, but made additions to his weekend getaway, a mansion on the Cahaba River named the Black Swann, which sported a fieldstone garage housing five sedans: two Cadillacs, two La Salles, and a Packard.

With economic destruction looming and obvious to everyone but himself, Swann decided to celebrate his accomplishments in January of 1931—the heart of the Depression—by leasing an entire railroad train and taking the University of Alabama football team, plus friends and business associates, to Pasadena for the Rose Bowl. The Pullman cars were stocked with libations. Professional chefs created daily menus of five-star quality with victuals in commensurate abundance. (The Alabama players would later comment that it was the best and most food they'd ever eaten.) After their 24–0 victory, Swann feted the team again. A line of black limousines picked up the players that evening and dropped them off at the Coconut Grove in Los Angeles, where they were entertained by one of the era's great stars, Will Rogers. Swann, of course, picked up the tab for everyone.

But Rome continued to burn while Swann fiddled. A little more than a year later, in 1932, he had to go hat in hand to his New York bankers for a loan to stay afloat. They accepted 156,000 shares of Swann Corporation stock—a majority interest—as collateral for a $750,000 loan. It was a more than generous gesture given the times and Swann's by now well-known extravagances. But the drained industrialist evidently couldn't make even the interest payments and the bankers called the note in 1933.

Rather than allow foreclosure by his bankers, Swann sought and got the help of two "white knights" from St. Louis, two fellow chemical industrialists, Edgar Queeny and Edward Mallinckrodt. The three hatched a plot that, while perhaps not criminal, certainly bordered on it. By setting up a series of complex stock transfers, along with the creation of holding companies and both straw and real corporations, the three conspirators somehow managed to cause the Swann Corporation, in effect, to disappear.

But then Swann's "white knights" morphed into sharks. As soon as the New Yorkers had been shoved out of the picture, Mallinckrodt took fast and good profits for his short-term investment and Queeny took over control of Swann's company—pushing Swann himself out with a million-dollar buyout. The stricken Swann retired to his drafty castle and apparently used the cash he received to pay off debts.

Queeny immediately renamed the company he'd acquired so dubiously. It was now to be called the Monsanto Corporation, and what a fine beginning it enjoyed. In less than a year, and during the lowest trough of the Depression, Queeny turned a stunning profit of nearly a quarter of a million dollars from Swann's Anniston operations.

One could reasonably conclude that Edgar Queeny's turnaround of the Swann Company was due not so much to his business acumen, but more likely to the fact that Swann could no longer squander company earnings on personal profligacy. Nonetheless, Edgar Queeny, a shark in knight's armor, would prove to be a remarkably able businessman for decades to come, turning Monsanto into the most successful agrichemical corporation in the world during his reign.

Ironically, Swann's financial demise and Queeny's gain would provide a windfall for his loyal workers, whom he had been paying with his "pass-offs." Queeny honored the pass-offs and then some. He paid Swann's employees with Monsanto stock, which had quickly risen to the giddy Depression-era price of $45 a share, with many of Swann's faithful employees becoming wealthy men while Swann began to slide into poverty.

Swann filed for bankruptcy in 1945 and was forced to sell his castle. Moving into a small, corner-lot house in a residential area of Birmingham, he could be seen tinkering with inventions on his screened porch for another decade. He came close to making a comeback too—patenting a specialized furnace for processing iron ore that he sold to Fulgencio Batista. But after a modest down payment to Swann, the Cuban dictator reneged on the deal. Reduced to selling aluminum window frames, the broken industrialist died on February 5, 1955.

Nobody knows how many tons of PCBs were produced by the Swann Chemical Corporation. But from 1930 to 1977, the year PCBs were banned by Congress, Monsanto allegedly manufactured more than half a million tons in its plants in Anniston, Alabama, and Sauget, Illinois. However, that amount was less than half the total production of PCBs by global chemical manufacturers. Perhaps a million and a half tons of PCBs, in large part produced under license from Monsanto, were eventually created worldwide.

What has already been released into the biosphere—about one third of the total production of PCBs—has thoroughly contaminated the planet and represents a deadly three-way threat to the biosphere. First, with high enough exposure, PCBs are lethally toxic. Second, PCBs can induce cancer in organisms after only minuscule exposure. Third, Swann's "magic fluid" puts entire species—and the planet's biodiversity—at risk because of its ability to disrupt the endocrine system, the complex, linked, hormonal factories that regulate growth and sexual reproduction in all vertebrates.

As far as humans are concerned, PCBs represent as much of a health mystery as tobacco did thirty years ago—which is no mystery at all. Although apologists for PCBs argue that nobody can prove, beyond any doubt, that there is a cause-and-effect relationship between PCBs and catastrophic human illness, scientists have known for decades that PCBs are lethal to human beings in even in minute amounts. Now they are finding out that in tens of millions of Americans, and hundreds of millions of genetically predisposed people globally, even infinitesimal body burdens of PCBs may trigger a wide range of disease—including serious immune system malfunctions, neurological disorders, birth defects, non-Hodgkin's lymphoma, testicular cancer, and a greatly increased risk of breast cancer in genetically susceptible women.

It can be said that Theodore Swann, self-taught chemist, dazzling industrialist, and tragic Southern gentleman, was the initiator of more chemically induced environmental destruction in the United States and the world than any other person in history, since he discovered how to commercially produce both PCBs and phosphates.

But what should also be kept in mind about the legacy of Theodore Swann is that the electrification of the country was the preeminent economic and cultural event of the twentieth century, and Swann's product—PCBs—was universally thought to be indispensable to the electrical industry. Believing that he was helping humanity, Swann didn't have an inkling of what his PCBs would do to the planet.

But Edgar Queeny, within a year or so of acquiring Swann's company, would start to understand the hypertoxic nature of the "magic fluid." Regardless, he and his Monsanto men forged forward. After all, good money had been paid out. Besides, making and selling PCBs was already a hugely profitable venture and would, Queeny believed, propel Monsanto to global prominence as a corporation.

Ed Queeny was correct. PCBs would do that for them. But shrewd as he was, Queeny must have known from the outset that he had made a Faustian bargain. So could the Monsanto chief and his executives weigh the profits against their growing knowledge that PCBs and their byproducts were a monster creation? Could they negotiate the problematic terrain presented by the need to hide the real character of PCBs? Yes, and then some.

CHAPTER 2

The Good Ol' Boys of Monsanto

The sublime accomplishment of technological civilization,
the comfort of Western industrial society, stands face-to-face with
a catastrophic destruction of the environment.

DR. ALBERT HOFMANN, INDUSTRIAL CHEMIST
AND DISCOVERER OF LSD

After the Great Chicago Fire of 1871 burned down his father's modest rental properties, John F. Queeny took the course taken by so many early American business tycoons: he dropped out of school to earn wages for his family. At the age of twelve, he got a job as an office boy at a drug company for $2.50 per week. He soon rose to delivery boy; twenty years later, through patient determination, the roughly handsome Irish-American became a full purchasing agent.

In 1894, John Queeny was recruited by the drug manufacturer Merck and Company and moved to New York City after being appointed sales manager. There he was to meet and marry the daughter of a Spanish aristocrat and Caribbean plantation manager. Olga Mendez Monsanto was not beautiful, but she was alluring, with thick dark-brown hair and luminous dark eyes, set off by a patrician aquiline nose. Slim and long-limbed, the elegant Spanish noblewoman, said to be a descendant of courtiers to Queen Isabella II, certainly made an interesting visual foil for her large-boned, craggy husband.

Again switching horses, Queeny moved with his new bride to St. Louis, where he became a purchasing agent for the Meyer Brothers Drug Company and Olga bore a son on September 29, 1897—Edgar Monsanto Queeny. While moonlighting, John Queeny started a small manufacturing company that produced saccharin, apparently ignoring the international patents held by the German chemical companies that had discovered the sweetener. After a series of major and minor mishaps (the original plant burned to the ground uninsured), Queeny started to turn a good profit, slickly guaranteed with long-term purchasing contracts from his creditors.

The budding industrialist probably wanted to give his new company his own name, but since he was still employed by the Meyer Brothers and moonlighting, it would have been awkward to do that. So Queeny chose his wife's maiden name, and the Monsanto Chemical Company was born. Queeny continued to build his enterprise by hiring three young Swiss chemists who were familiar with European manufacturing processes, and Monsanto was soon producing caffeine and vanillin (the main synthetic ingredient of vanilla) along with saccharin. By 1920, Monsanto was continuing to grow and would do so—albeit fitfully—until Queeny's son and heir took over as president in 1928.

By his early thirties, Edgar Queeny was already a star—an odd and brilliant one—who would far outstrip his father in accomplishment and also symbolize the dark, take-no-prisoners corporate culture at Monsanto for three decades. Lean, tall, and handsome, he had the mildly swarthy looks of a young Spanish grandee; he inherited his mother's nose and her elegance and none of his father's Irish roughness—at least physically.

But the son did share some of the father's tendency toward excess, though not necessarily in exactly the same flavors. The elder Queeny smoked cigars incessantly. Edgar chain-smoked cigarettes. The father was known to like more than his share of bourbon. The son had a similarly large appetite for alcohol, but preferred the best scotch and dry martinis. Their personalities were as dissimilar as their tastes. John Queeny was an Irish charmer with a volatile temper that would subside as quickly as it erupted—a classic extrovert—while his son was an almost sociopathic introvert who would offer no apologies for a savage rudeness that allegedly never softened in his lifetime.

At executive meetings, Edgar Queeny was famous and feared for his pointed and extravagant yawns. If the hapless underling didn't get the message, then Queeny's head would nod as if he were going to sleep. Edgar would also utilize a stony silence with people who bored him. There were many meetings in which Queeny wouldn't say a single word. After fidgeting and repeatedly putting his legs up on his desk or nearby furniture, Queeny might simply leave the room, never uttering a sound, with the other person in mid-sentence.

But Edgar's bizarre behavior had no bearing on his abilities, as both a superlative businessman and creative intellect. During leaves of absence from his duties as chief of Monsanto, Edgar wrote a bestselling business book (essentially an anti–New Deal diatribe) and another that was recognized by experts as *the* classic book on duck hunting. He also filmed and produced numerous documentaries at his film studios in St. Louis. Subjects ranged from salmon fishing (purchased by Warner Brothers as a movie "short" in the 1950s) to anthropological studies of African tribes sponsored by the American Museum of Natural History.

Fine as they were, Edgar Queeny's cultural and intellectual accomplishments never matched the level of success that his predatory business expertise would produce. By the beginning of 1934, Queeny had completely subsumed Swann Chemical into Monsanto. Not only did he successfully "hide" the company from the East Coast bankers who had been quasi-swindled, Queeny paid off his partner, Ed Mallinckrodt, and gave Theodore Swann a cool million in hard, Depression-era cash for his company in the takeover. But the showpiece of Queeny's corporate legerdemain was the turnaround he accomplished financially. From an investment of a little over two-and-a-half million dollars, Queeny was reaping profits approaching $500,000 within three years of purchasing Swann Chemical, all in the depths of America's worst depression.

But as Monsanto flourished like a lone sunflower in the stunted fields of American business, its seeds were carrying a toxic chemical genotype. In the late spring of 1934, workers at Monsanto's Anniston, Alabama, plant came down with chloracne—a serious, disfiguring skin disease that appeared to be related to the PCB manufacturing process. Chloracne was first identified in German industrial workers in 1897. At the time, scientists thought that the disease was associated with exposure to toxic chlorine, thus it was given the name "chloracne." The symptoms included eruptions of large blackheads, hard white cysts and pustules that would cover the worker's face, ears, armpits, and groin. Even if the case of chloracne was only moderate, it would still leave permanent, disfiguring scars wherever the skin eruptions took place.

The chloracne outbreak experienced by the workers in 1934 at the Anniston plant was serious enough that Monsanto hired a physician, Dr. Frederick B. Flinn, to conduct a series of tests using rabbits as subjects. Nineteen hapless animals were shaved, and patches containing Aroclor—the new commercial name Monsanto gave to its PCBs—were applied to their skin. Within days the PCBs had produced ulcerous lesions on the test rabbits.

While Dr. Flinn tested, the situation at Anniston was getting worse—much worse. Workers operating the Aroclor distillers were coming down with severe cases of chloracne and they were doing the unthinkable by *suing* the company for disfiguring them. With Monsanto profits at risk from settlements and court judgments, Queeny ordered quick action. Basic ventilation was built into the Aroclor manufacturing area for the first time. Workers there were furnished with a complete set of clothing each day, and provided with cold cream to apply to their faces, arms, and necks. Baths were built at the plant and Aroclor workers were required to bathe after their shift and rub themselves down with alcohol (on their own time).

Monsanto was to claim that these actions were all that was needed to make the handling of PCBs a completely safe enterprise, and in fact, cases of chloracne at Anniston were reduced. But Monsanto never bothered to alert its customers that handling and using PCBs was dangerous unless rigorous safety measures were put in place. It was to be a fatal omission.

In 1936, around Christmastime, a healthy twenty-one-year-old worker in a large New York factory that manufactured electrical wiring complained of being badly constipated as well as having severe abdominal pain and bloating. On being admitted to the hospital, he was jaundiced and apparently suffering from advanced liver disease. His arms, face, chest, and back were covered with pustules and large blackheads. He died shortly after being examined by doctors.

Within weeks another worker from the same plant, owned by the Halowax Corporation, working in the same area—an electrical-wire-coating facility that used heated Aroclor for insulating electrical wire—was taken to the hospital with acute jaundice. He too died shortly after admission. His best friend at the factory also was hospitalized with jaundice and died within two weeks. In the case of the third fatality, a detailed autopsy was performed and it was found that the healthy young man had died of "acute yellow atrophy" of the liver—with no preceding attacks of jaundice or any other significant health problems.

Sanford Brown, the president of Halowax, seeing potentially serious problems ahead for his booming company, hired a renowned Harvard professor to study not only the deaths of the three workers but the illnesses of dozens more who were sickened but did not die. Dr. Cecil Drinker, an occupational health and safety expert, known for thoroughness and insightful research, quickly concluded that the toxic substance that had produced the fatalities and widespread sickness among the Halowax employees was none other than Theodore Swann's "magic fluid." All the handlers who died had one thing in common: they had been exposed to hot Aroclor.

At his Harvard laboratory, Dr. Drinker performed tests on the PCB samples he had received, this time using both rabbits and rats in order to give a better picture of cross-species mammalian toxicity. His findings were clear and impossible to misinterpret: PCBs were very toxic substances, even at relatively low levels of exposure. They attacked the liver, causing not only observable damage to the organ but death to animals exposed to vapors at levels well below what workers would normally encounter on the factory floor. Drinker also found another facet of the toxic nature of PCBs that would have major consequences for the global environment and make scientific research into the "magic fluid" a difficult proposition. Dr. Drinker was the first scientist to discover that the various Aroclors affected test animals in differing ways—some being much more toxic than others.

(Monsanto manufactured a number of different blends of PCBs, Aroclors with varying amounts of chlorine: Aroclor 1236, Aroclor 1242, Aroclor 1248, Aroclor 1252, Aroclor 1254, Aroclor 1262, Aroclor 1269, among others. The names were based on chlorine content in the mixture, with the last two digits indicating the percentage of chlorine. For example, Aroclor 1236 has 36 percent chlorine content; Aroclor 1254 has 54 percent; and so on. Generally, the higher the chlorine content, the longer the PCB stayed in the environment, with the highest-chlorine Aroclors being virtually indestructible via natural biodegradation. The most toxic Aroclor that Monsanto sold, Aroclor 1254, was also its most popular PCB formula. Perhaps every person reading these words has molecules of Aroclor 1254 in his or her tissues and/or blood.)

With the Drinker report complete in 1937, Halowax president Sanford Brown called for a summit conference on the PCB problem. Invitations were sent to Monsanto and some of its other large customers, such as General Electric. Concerned state health officials, staff scientists, and Dr. Cecil Drinker and his associates at Harvard were invited as well. It would be the first, last, and most illuminating meeting of its kind on the "magic fluid."

The first speaker at the conference, Dr. W. R. Oettingen, an industrial toxicologist from Maryland, complimented Sanford Brown for having the courage to confront the PCB issue. He then went on to praise Cecil Drinker for doing cross-species testing of PCBs since "human testing" was out of the question. Next, the director of occupational hygiene for the state of Massachusetts, Dr. Manfred Bowditch, voiced a serious practical problem—the simple solution of which would be resisted by Monsanto for decades.

Dr. Bowditch wanted Monsanto to label its products as toxic. He stated that earlier in the year he had issued a "strongly worded warning letter" on the lethal toxicity of PCBs to all the manufacturers using Monsanto's PCBs in Massachusetts. However, he didn't think his notice would be effective because Monsanto only used an esoteric numbering system (the Aroclor numbers explained above) to designate their products that contained PCBs, with no warnings about toxicity or hazards to handlers.

However, the worst problems with PCBs revealed at the summit conference were those at the General Electric Wire Works plant in York, Pennsylvania. The plant manager, F. R. Kaimer, said, "We had in the neighborhood of fifty to sixty men afflicted with various degrees of this chloracne about which you well know. Eight of the ten of them were very severely afflicted, horrible specimens as far as their skin condition was concerned." Kaimer went on to relate that they had had one fatality, and although it couldn't be proven, it was almost certainly caused by exposure to Monsanto's Aroclors.

Kaimer said that his first reaction "was to throw it out—get rid of it from our plant. We didn't want anything like that for treating wire. But that was more easily said than done." Kaimer continued, "We might have just as well said that we will close our business and throw it to the four winds, because there was no substitute. And there is none today in spite of all the efforts we have made in our own research laboratories to find one."

Next at the conference came the turn of the man who would be the éminence grise of Monsanto for the next thirty years when it came to the "magic fluid," Dr. R. Emmett Kelly. A physician, Kelly would not only craft Monsanto's political policy regarding PCBs, he would deliver his unfailing defense of their Aroclors both publicly and inside the corporation. The conference attendees were to hear what would become a classic Kelly speech, variations of which he would deploy with great success over his long career at Monsanto. "I can't contribute anything to the laboratory studies, but there has been quite a little human experimentation in the last several years, especially at our plants, where we had been manufacturing this chlorinated biphenyl," Kelly stated. Then, with a bland delivery hiding a brazen lie, the physician said, "It has been our observation, that although on one occasion we did have a more or less extensive series of skin eruptions, we were never able to attribute as to a cause, whether it was impurity in the benzene we were using, or to the chlorinated biphenyl."

Dr. Kelly continued: "We have never had any systemic reactions at all in our men. We have examined them very closely both from what laboratory tests we thought might help us, and from the clinical viewpoint. Also from the chlorinated biphenyl alone, there had been no cases of systemic poisoning reported. I don't believe that we can transpose the laboratory results into the actual humans, not paying considerable attention to the volatility of different substances and the way they are being used."

Yes indeed, there had been "quite a bit of human experimentation" among the trusty and sturdy workmen that Theodore Swann had recruited to manufacture PCBs. And the fact was that these unsuspecting human guinea pigs consistently demonstrated that PCBs were *highly* toxic to some of them—if not the majority of workers exposed to Aroclor fumes. This was why these loyal workers reluctantly sued Monsanto and why Monsanto was forced to institute a plan for dealing with PCBs that included an exacting hygiene regimen at their Anniston plant.

And when dissembling facts was not necessary, Dr. Kelly would nimbly dance around the truth with statements like "from the chlorinated biphenyl alone, there had been no cases of systemic poisoning reported." The key word there was "alone." Dr. Kelly wasn't saying that there *hadn't been* cases of systemic poisoning—because indeed there had been, since chloracne is a symptom of systemic poisoning in its own right. What Kelly was really saying was that nobody could *prove* that the systemic poisoning of the workers was caused by the PCBs alone.

In an uncanny parallel to the position of the tobacco industry some fifty years later, Dr. Kelly's statements at the PCBs summit were to be the philosophical underpinning of Monsanto's decades-long defense of PCBs: you couldn't *prove* that PCBs caused harm to humans because there were always other toxic substances that could be blamed. (Some of these other toxic substances, such as dioxins, were actually byproducts of the PCB manufacturing process itself.)

But the last point in Dr. Kelly's statement was perhaps the most disingenuous and effective. When Kelly said that "I don't believe that we can transpose the laboratory results into the actual humans," he was basically charging that all Dr. Cecil Drinker's painstaking work and that of his own colleague, Dr. Flinn, was essentially worthless since lab animals weren't human beings.

Of course the truth was that scientists had been successfully using animal subjects as substitutes for humans routinely for hundreds of years before Dr. Kelly's statement. They had to because the ethical bedrock of medicine, that the physician should first and foremost "do no harm," prohibited the involuntary testing of toxic substances on human subjects. But Dr. Kelly, as he so cavalierly made clear, had no qualms as a physician about using Monsanto workers as research guinea pigs. (In fact, some might find Kelly's thinking similar to that of the medical researchers of the Third Reich: people actually made the *best* test subjects, since the data could be applied across human populations. In other words, when you have people, why use rabbits?)

In any case, nobody raised any questions about Dr. Kelly's statements. Halowax's president, Sanford Brown, closed the meeting on a note of semi-paranoia, warning the group, especially the state health officials, that they should be careful not to create "mob hysteria on the part of the workmen in the plants" regarding PCBs. One could almost hear a depressingly long sigh preceding Brown's final words in the transcript of the conference. "This thing may continue, probably will continue, for years."

CHAPTER 3

The Long Con

*If we continue to devote our chief energies to further extending
our commerce and our wealth, the evils which necessarily accompany
these, when too eagerly pursued, may increase to such gigantic
dimensions as to be beyond our power to alleviate.*

ALFRED RUSSEL WALLACE, *The Malay Archipelago*, 1869

In their 1944 salesman's brochure, after five pages of descriptions of
their product, Monsanto finally got around to the toxicity of PCBs.
There it was for the first time: a declarative statement that Aroclors were
toxic. The company brochure then went on to describe the symptoms
of PCB poisoning as primarily chloracne and "acute yellow atrophy of
the liver." But there was a carefully couched, contradictory kicker. The
toxicity and symptoms, the brochure said, applied only to the "animal
organism."

Dr. Emmett Kelly had apparently changed his mind since 1937,
when he was of the opinion that testing PCBs on animals was essentially worthless. But apparently Kelly still could not allow himself to
let anyone at Monsanto admit that PCBs were toxic to human beings.
This, even though he knew of more than a decade of incontrovertible
evidence to the contrary—including the deaths of at least three workers
and the poisoning of hundreds, perhaps many thousands, more (including servicemen and women, and thousands of defense workers during

World War II who handled PCBs without any precautions, given the exigencies of the era).

While the Aroclor drummers at Monsanto may have neglected to add human safety data for their product, they certainly didn't forget to extol its virtues. From Monsanto promotional materials, it appeared that Aroclors were indeed a "magic fluid" that could make a veritable rainbow of products irresistibly durable and useful. Coating electrical cables and wiring with the product would render them nonflammable in case of short circuits where arcing might otherwise occur and start fires. Because PCBs could resist breakdown from prolonged exposure to high temperatures, the Monsanto marketers pushed their Aroclor as an unsurpassed hydraulic fluid for use with critical components in heavy equipment such as brake cylinders and control governors.

And there were more prosaic uses for Aroclor in domestic markets. Monsanto exalted its product as the perfect component for all varieties of paints and lacquers. According to the Monsanto chemists, adding PCBs to enamels (specifically the most toxic, Aroclor 1254) would not only give the paint terrific adhesive qualities, but would make the surface as hard and pretty as the most expensive shellac. For lacquer paints, Monsanto advised potential customers that its heavier Aroclors would give lacquers a high permanence and "extreme chemical stability," allowing house paints to have "remarkable durability" through the addition of PCBs.

(It must be noted here that, from their own extensive testing, Monsanto understood the extreme dangers that PCBs posed when used in paints. Monsanto scientists were keenly aware that small concentrations of Aroclor vapor could be dangerous and possibly even fatal unless great care was taken to ensure proper ventilation when paints were applied. There is no record that Monsanto ever warned unsuspecting wholesale buyers or consumers of the toxic dangers posed by PCB-laden paints.)

Monsanto's hyping of its Aroclor line reached a peak with two applications that today might have left Monsanto open to major lawsuits, if not criminal negligence proceedings. The first was the suggested use of Aroclors for impregnating protective clothing for firemen, given the heat-resistant characteristics of the "magic fluid." But the problem with this concept was that all appropriate testing of PCBs showed that they became far more toxic—lethally toxic—when heated. Not only that, the firemen would be continuously breathing the Aroclor vapor under the worst conditions: confined, hot, and smoke filled. There are no available records showing that Monsanto actually sold Aroclor for this use; conversely, there is no public data confirming that they didn't.

In all likelihood, the most bizarre and worst Aroclor use conceived by Monsanto's marketing brain trust was application of the chemical directly on fishing-pole handles. Monsanto scientists went so far as to run tests on hickory fishing-pole handles to determine just how to obtain the best adhesion results. They found that the manufacturers of the fishing poles permeated with Aroclor would need to apply the PCB formula at around 200 degrees Fahrenheit—a temperature at which the Aroclor fumes killed test animals within hours and were almost certain to cause severe liver damage to any workers in the area. And in terms of direct contact with PCBs, with the fishing pole in hand for hours, if the unsuspecting anglers didn't wear gloves, they might absorb enough PCBs to cause permanent systemic damage while waiting for that lunker bass to strike.

While Dr. Emmett Kelly may not have actively participated in the product hyping of the Aroclor line, he was always available and willing to help gloss over their toxic aspects—and those efforts left him a busy man at Monsanto. There were so many complaints and queries regarding the poisoning of unsuspecting workers that Dr. Kelly had developed a cookie-cutter response by the early 1950s—a response that was quite effective in deflecting any responsibility away from Monsanto's doorstep.

Dr. Kelly would invariably inform the customer/complainant that Aroclors were being used in a way that was "contrary to our expressed instructions." Monsanto couldn't be held responsible if, for example, PCBs leaked from a transformer in a warm, enclosed area and the vapors made maintenance workers sick—its Aroclors couldn't be blamed for causing illness if they were inappropriately deployed.

The next part of Kelly's set response to complaints about Aroclors was to generally obfuscate the established research on PCBs. A typical example of this strategy could be found in a letter Kelly sent to the head of Indiana's State Board of Health. Kelly wrote, "The toxicology of aroclors is somewhat confused. The experimental work done by Dr. Cecil Drinker at Harvard was done in connection with PCBs. In that particular work at Harvard, Dr. Drinker found that aroclors 1268, which means it is chlorinated to sixty-eight percent, was of low toxicity. The confusion existed in his findings that aroclors 1254, which is chlorinated to fifty-four percent was considerably more toxic on inhalation. We did not supply him with this material, and I was never convinced that some error might not have been made in the sample."

Put differently, the credibility of Dr. Cecil Drinker, one of America's leading industrial toxicologists, whose state-of-the-art research with PCBs had never been refuted, was maligned by Dr. Kelly because Monsanto didn't supply Dr. Drinker with the samples of PCBs used for the lab tests. Of course, Kelly didn't bother to inform the worried state health chief that the samples Drinker used were supplied to him by the Halowax Corporation, and they—like every other corporate purchaser of Aroclors in America—bought their product from Monsanto.

In 1952, Monsanto was again having problems with the labeling of PCBs. During the PCB summit conference almost twenty years earlier, Monsanto was beseeched by the director of the Massachusetts Division of Occupational Hygiene to more clearly label their Aroclors so that workers would know they were handling highly toxic and potentially deadly materials that should be used only under strictly controlled situations. Monsanto had ignored this appeal, but now there was pressure coming from the federal government that wasn't going to be easily diverted.

At the request of the U.S. Public Health Service, Monsanto reluctantly agreed to label PCBs with the mild and easily misinterpreted statement: AVOID REPEATED CONTACT WITH SKIN AND INHALATION OF THE FUMES AND DUSTS. Even though they were loath to admit to any potential human health problems with their popular and highly profitable product, at least one Monsanto executive saw a benefit for the company in being forced to label their PCBs as dangerous. He wrote in a memo that "the very few instances when Aroclor/PCBs have been misused, especially at elevated temperatures, it has led to lawsuits. [Therefore] it is highly desirable and protective to us in having our current label on the Aroclor packages."

The executive went on to joke about how far they'd come from the bad old days. He wrote, "Back in 1936 or thereabouts, when the Aroclor applications were relatively few, and the customers about equally few, there was indeed the prize application using Aroclor 1254 as a chewing gum plasticizer." He would add, ironically, "The wording of our label would not be compatible with this sort of thing."

A few years later, queries about PCB toxicity came from a potentially huge customer. None other than competitor and fellow chemical industry giant Dow Chemical wanted more information about PCBs. Evidently interested in Aroclors because of all the product hype, Dow told Monsanto that they intended to use its Aroclors as a plasticizer in a new product meant for the home market—Saran Wrap. Saran Wrap was to be a clear plastic wrap that Dow wanted housewives to buy in order to store food and to wrap sandwiches for the lunch boxes of their children and husbands. But Dow's chemists had heard rumors about workers getting sick from Aroclor fumes and they wanted assurances from Monsanto that the product was completely safe.

The request triggered a "confidential" reinvestigation of PCBs by Monsanto. The goal was simple enough, according to an in-house memo: gather new information "in an attempt to convince Dow that the use of Aroclor in their plant was safe." The tests that Monsanto ran were "air concentration" tests. They used PCB-impregnated paints in rooms of different sizes with different levels of ventilation. The results were mixed, to say the least. The Monsanto sales department was informed that "The conditions of the tests and the limitations of the apparatus are such that absolute determinations . . . are not possible." And further, Monsanto's own scientists who prepared the Saran Wrap report refused to endorse the safety of Aroclors where human exposure was involved. "Any recommendations on the lack of toxic effects of Aroclor/PCBs in the concentrations found will have to come from the Medical Department," said Dr. Emmett Kelly.

There is no record that Dow ever used PCBs as plasticizers in Saran Wrap. However, Monsanto's "air concentration" tests raised another worker safety issue that was difficult to explain. According to Monsanto documents from the 1950s, the company applied very different safety standards regarding PCBs at their two main manufacturing facilities, which they apparently didn't intend to pass along to Dow Chemical.

At their St. Louis–area plant, the Krummrich factory, the entire Aroclor production building, was "rated a toxic department." All workers were provided with clean sets of clothing—hat, shoes, coat, trousers, underwear, socks, and rubber booties—for each shift. The men were required to bathe after work and it was on "paid time." No eating was allowed in the PCBs manufacturing building and workers in the Aroclor department were given an annual medical examination and lung X-rays every three years.

How different things were at the Anniston, Alabama, plant, where there many less educated African American workers manufacturing PCBs—handling the very same toxic materials that the St. Louis workers did. At Anniston, no special protective clothing was issued to the Aroclor operators. According to Monsanto records, "A daily change of clothing was provided in the past, but this practice ceased before the war." Tins of cold cream ointment were made available, but each worker made his own decision about whether to protect his skin. As far as bathing after work was concerned, "a good quality soap and alcohol for rubbing down purposes are provided. The men are expected to take a bath in their own time at the end of the shift. The operators are sufficiently trained in the need for personal cleanliness, that a record of bath taking is not warranted."

Monsanto—after more than twenty years of convincing customers of the safety of PCBs—finally came up against a truly tough and skeptical customer: Admiral Hyman G. Rickover of the United States Navy. Admiral Rickover was the genius father of America's most advanced military technology of the time, nuclear submarines. He was brilliant, abrupt, and a stickler for guarding the health of his submariner crews. Rickover and his engineers were looking for a substance to use in the hydraulics of submarine periscopes, and Monsanto marketers had hyped their Aroclor line as the answer to their needs. But Rickover and his men were leery.

Unlike Monsanto's corporate customers, the admiral ordered a series of tests on PCBs in the mid-1950s to be done by the famed physician and scientist Dr. Albert Behnke at the Naval Medical Research Institute. The Behnke tests turned out to be a disaster for Monsanto. Skin applications of Monsanto Aroclor mixtures on rabbits killed them all. Animal inhalation tests at relatively low levels caused "definite liver damage."

Appalled that Monsanto had even suggested using PCBs on his nuclear subs, Rickover wanted nothing to do with them or their products. Dr. Emmett Kelly was miffed at the admiral's curt rejection of their Aroclors as being inappropriate for use in nuclear submarines. Kelly knew if it got around that Monsanto's PCBs had been banned in the Navy's submarine fleet, it couldn't do anything but hurt sales. So with an industrial chemist tagging along, Dr. Kelly flew to Washington to confront Rickover's picky naval procurement people and convince them that PCBs were fine for their nuclear submarines.

Dr. Kelly told the navy men that the concentrations of PCBs would never reach toxic levels in a submarine. Although the naval officers' reaction to Kelly is not known, they were evidently less than impressed with the Monsanto physician and his opinions. "No matter how we discussed the situation, it was impossible to change their thinking," Kelly complained to a colleague after the meeting. "The Navy does not appear even willing to put Aroclors in a trial run to see if it will work."

Rickover's rejection of PCBs may well have caused Monsanto additional credibility problems. Approximately a year later, in 1958, Mobil Oil demanded that Monsanto label PCBs sold to them with a "caution stamp." Mobil's command evidently caused a wave of consternation at Monsanto. Caught between the ever-present conflict of balancing the chance of lawsuits against sales, Monsanto had been wrestling with its labeling problem since Edgar Queeny started the company back in the early 1930s. PCBs were robust sellers internationally, and labeling them

as toxic would certainly scare off some big buyers. However, by *not* labeling them as potentially dangerous, Monsanto was left vulnerable to
lawsuits by both customers and workers.

In response to the Mobil requirements, Monsanto behavior would
once again skirt the realm of criminal negligence given their knowledge
of the lethality of their Aroclors. "It is our desire to comply . . ." wrote a
Monsanto executive in a secret memo. "But to comply with a minimum
and not give any unnecessary information which could very well damage our sales position in the synthetic hydraulic fluid field."

But Monsanto was to resolve its chronic Aroclor/PCB labeling problem a few years later with the ultimate evasion of responsibility. Because
Monsanto sold Aroclor in so many varieties, and since each variety had
different toxicity characteristics, Monsanto informed all purchasers of
Aroclor that they were instituting a new policy regarding PCBs: "The
ultimate responsibility for the proper labeling of a formulation remains
with the customer."

There was no official reaction from Aroclor buyers, small or large,
to Monsanto's pronouncement. But judging from Monsanto's Aroclor
sales figures for the 1950s and early 1960s, their clients accepted their
new responsibility for the toxicity of the "magic fluid" without com-
plaint.

But there were exceptions. A furious chief engineer at a chemical
company in the Bronx, New York—Hexagon Laboratories—wrote to
Monsanto's Dr. Kelly that he had two very sick workers, one of whom
almost died from inhaling Aroclor 1248. There wasn't any doubt in the
mind of the Hexagon engineer that labeling by Monsanto was negli-
gent. "Since we are dealing with a highly toxic material at high temperatures," he said, "it is felt that a more thorough and clearly written
description of the hazards be described under 'safety and handling.'"
He also denounced Kelly for not including antidotes or first-aid treatment on Aroclor labels for workers exposed in an accident. (But this
last request Dr. Kelly couldn't meet if he had wanted to, as there were
no known antidotes for PCB poisoning, and there still aren't, as we
will see.)

With his usual nonchalance, Kelly advised the Hexagon chief engineer to give him a call "or have a doctor call me if they needed any further information." However, Dr. Kelly's in-house response to the Hexagon Laboratories poisoning was cold-blooded and couched in barely veiled racial terms. Dr. Kelly reported to his superiors at Monsanto that "one individual was under the care of the physician, and the physician suspected liver damage, although no jaundice could be seen. The patient was a Negro and was not hospitalized."

The darker complexions of some African Americans could mask noticeable skin yellowing—a main symptom of jaundice and liver toxification. The Monsanto medical director and physician certainly must have known that jaundice also yellows the whites of a victim's eyes and is easily diagnosed that way. (There is an untoward ending to the Hexagon PCBs story. According to published reports, the owners of Hexagon Laboratories abandoned the facilities in 1988, leaving behind tons of toxic waste, including Monsanto PCBs in leaking barrels and underground chemical storage tanks. Hexagon Laboratories property was declared a New York State Superfund site and, in the early 1990s, a $1.2 million cleanup was initiated. But as of late 2002, the area was still highly contaminated, according to community activists in the Bronx.)

By the early 1960s, Dr. Emmett Kelly had probably dealt with hundreds of queries and complaints about the toxicity of Monsanto's Aroclor line. For almost thirty years, Kelly had known that PCBs were so lethal that they could be fatal if inhaled under certain circumstances. He also knew that Aroclors could be highly toxic to the liver, causing irreversible damage and death in even comparatively low concentrations. And Kelly surely had known of thousands of cases of severe, disfiguring chloracne induced by PCBs.

Yet, when another physician, Dr. Marcus Key, an administrator with the U.S. Public Health Service, asked Emmett Kelly, doctor and supposed healer, about the dangers that Aroclor posed to humans, Kelly responded with a reply that would have pleased Joseph Goebbels. "As I told you on the telephone, our experience and the experience of our customers over a period of nearly twenty-five years, has been singularly free of difficulties. To our knowledge, there have been only three instances where chloracne has occurred. In view of the millions of pounds which had been produced and used in very many applications, the low frequency of any difficulties has been gratifying."

CHAPTER 4

The Discovery

It was to be an event that would change human consciousness forever. On June 16, 1962, the *New Yorker* magazine began a three-part series by a pop science writer named Rachel Carson. Using scientific studies based on findings made newly possible by electron capture detectors, the fifty-four-year-old Carson warned that Earth's environment was being poisoned by synthetic chemicals, primarily the insecticide DDT.

Three months later, Houghton Mifflin published Carson's complete work on the subject in a book titled *Silent Spring*. With punctilious documentation and lissome prose, it quickly became a bestseller. The heart of the book was captured in its last paragraph, where Carson wrote, "The 'control of nature' is a phrase conceived in arrogance, born of the Neanderthal age of biology and philosophy, when it was supposed that nature exists for the convenience of man. The concepts and practices of applied entomology for the most part date from that Stone Age of science. It is our alarming misfortune that so primitive a science has armed itself with the most modern and terrible weapons, and that in turning them against the insects, it has also turned them against the earth."

While Rachel Carson's words apparently struck an innately rational chord in her readers, they also caused chemical manufacturers to seethe publicly against the "foolish cultist." And given the times, as one might expect, some of the attacks were over-the-top sexist. "I thought she was a spinster," one critic was said to have remarked about Carson, who was by then dying from breast cancer. "What's she so worried about genetics for?" Another fellow scientist and sometime industry spokesman was hardly less caustic, writing that "the major claims of Miss Rachel Carson's book . . . are gross distortions of actual facts, completely unsupported by scientific, experimental evidence and general practical experience in the field."

Certainly an attack on their values by a misguided spinster wasn't going to sit well with Edgar Queeny's men at Monsanto. Shortly after Carson's book was published, the Monsanto public relations machine went into action, printing and distributing five thousand copies of a their own mini-book titled *The Desolate Year*, a parody of *Silent Spring* that began with an apocalyptic description of Earth stripped bare by voracious insect hordes.

Apparently the Monsanto campaign to discredit Carson was successful—at least with some mainstream journalistic outlets. *Newsweek* casually dismissed the importance of *Silent Spring*, saying that federal studies of pesticides were under way and that there had already been recommendations for controls as well as information on the dangers of misuse. *Time* magazine, then dominated by the unbridled pro-business views of founder Henry Luce, carried a fierce attack on Carson's credibility. "Despite her scientific training, she rejected facts that weakened her case, while using almost any material, regardless of authenticity, that seemed to support her thesis."

It was also a time when, besides comparatively rampant sexism, cash from large corporations for slanting and/or planting stories was not unheard of at some publications. This is not necessarily to impugn the integrity of any journalists of the era at either *Time* or *Newsweek*, but forty years later, *Time* would change its position considerably, naming Rachel Carson as one of the one hundred most influential people of the twentieth century. *Time* editors were to write that "all but the most self-serving of Carson's attackers were backing rapidly toward safer ground. In their ugly campaign to reduce a brave scientist's protest to a matter of public relations, the chemical interests had only increased public awareness."

But vindication was never to be in Rachel Carson's lifetime. Her diagnosis of terminal breast cancer came while she was approximately halfway through writing *Silent Spring*. Carson died two years later at the age of fifty-six. President Jimmy Carter posthumously awarded her the highest civilian honor that can be bestowed, the Presidential Medal of Freedom, in 1980. But her legacy as eco-heroine was perhaps not as important as her role as gentle thrower-down of gauntlets to her fellow scientists. Carson's informed speculation was that not only were DDT and other pesticides contaminating the American environment—they almost certainly must be wreaking similar havoc on ecosystems throughout the globe.

In Sweden, based on Carson's call to arms, the Swedish Royal Commission on Natural Resources ordered an immediate chemical survey in regions of the country thought to be affected. The commissioners wanted to see if DDT was present in the Swedish environment, and for the task of finding out they picked the department of analytical chemistry at the University of Stockholm.

There the research fell to a young academic, a chemist named Soren Jensen. His department had recently acquired an electron capture detector, and with his powerful new tool, the grad student promptly started taking samples from all around the Swedish countryside to see how bad the DDT contamination might be. Soren Jensen knew from Carson's book, as well as other scientific literature, that birds were highly sensitive to DDT. Consequently he started his research by analyzing a mature white-tailed eagle that had turned up dead from unknown causes. Jensen carefully prepared tissue extracted from the bird's fat because DDT was known to be lipophilic (meaning it liked to concentrate in fatty tissue).

The first chromatogram the young chemist did showed that the specimen eagle was indeed contaminated with DDT and a metabolite of the pesticide called DDE. But the graph clearly showed fourteen other peaks besides those for DDT and DDE. It was very odd. At first,

Jensen thought he had made some sort of mistake. Perhaps his sample was contaminated, or maybe his electron capture detector was out of whack since these already had a reputation for being finicky instruments at times. To double-check his techniques, Jensen used a pike—a kind of small freshwater barracuda that had been caught in a remote Swedish lake. Again, the chromatograms for the pike showed the strange fourteen-peak signature.

Stumped, Jensen searched all the scientific literature he could find and still came up with nothing. Whatever the mysterious molecules were, they hadn't yet been identified by scientists as an environmental contaminant. As he was to write a few years later—after his discovery had rocked the global scientific establishment—"the possibility of a pure analytical solution of the problem was, at that point, impossible."

Researchers stuck with a scientific whodunit use the tried-and-true methods of homicide detectives: they exclude suspects, and this is precisely what the dogged Jensen began to do. The first thing he needed to establish was that the unknown substance was not simply a "common natural product." To determine this, Jensen used an uncomplicated but elegant approach. He had the Swedish Museum of Natural History collect more pike, this time from a group of secluded, pristine lakes in Lapland. The new samples showed less of the mystery chemical, indicating that it was probably a pollutant somehow related to human activity.

Jensen then asked himself another question with important exclusionary value: Just when did the mystery substance start to pollute the Swedish environment? To find this out, the resourceful researcher went back to the Swedish Museum of Natural History and obtained a single feather from each white-tailed eagle in the museum's collection, starting in 1888.

Then things started to jell. Jensen discovered that the unknown substance was measurably present for the first time in 1942. This meant that it couldn't be any type of industrial pesticide or byproduct, since they weren't used in Sweden until after 1945. Jensen had also found the odd chromatogram signature in just about all levels of the marine food chain, including salmon, and in other top predators like eagles. He also discovered that the mysterious substance seemed to magnify up the pyramid of prey and predator. And according to the response times on the gas chromatograms, the compound was highly stable, perhaps uniquely stable, indicating it was going to stay in the environment for many, many years.

Still Jensen was stymied. Having excluded nearly all that could be reasonably excluded, he still didn't have the name of the omnipresent ecological trespasser. But as the Hollywood homicide investigator often gets lucky, so did Soren Jensen. His good fortune came in the form of another white-tailed eagle found dead in the Stockholm archipelago. This bird "contained enormous amounts of the unknown substance," wrote Jensen in his journal. It was so loaded with the pollutant that Jensen was able to run a series of analytical tests that could only be performed if there were extremely high levels in the eagle's tissue. The results indicated that chlorine was present in the sample. By postulating the amount of chlorine in each molecule of the material in relation to the possible molecular weight, Jensen was able to arrive at a plausible formula for biphenyls that were polychlorinated—PCBs.

Jensen quickly verified his discovery. He took laboratory test samples of chlorinated biphenyl and ran them through an "extensive gas chromatographic investigation." The peaks on every single graph produced by the commercial PCBs matched the unknown peaks from the white-tailed eagle and all the other animal species that Jensen had found to be contaminated. Not only had Jensen discovered PCB contamination, he realized that some of the peaks represented different kinds of PCBs, and this was what had been making the substance so difficult for him to analyze. There wasn't just *one* PCB in the Swedish environment—there were *many* kinds of PCBs out there.

Even more dizzying for Soren Jensen was the good probability that much of the research used by Rachel Carson in *Silent Spring* was distorted by the undetected presence of PCBs. Since Jensen was the first to discover ubiquitous PCB contamination, the DDT researchers in the United States and in Britain had apparently accepted that the odd peaks they surely must have found in their analyses were either DDT metabolites or measurement anomalies . . . when in reality they were finding varieties of Swann's "magic fluid."

• • •

In its December 1966 issue, the prestigious British journal *New Scientist*
announced that the planet was apparently being poisoned by a toxic industrial chemical that had inadvertently been allowed to escape into the
environment. Titled "A New Chemical Hazard," the item was just three
paragraphs long and mentioned Soren Jensen only in passing. The editors of *New Scientist* couldn't have had an inkling that their blurb about
a young chemist's analytical work—shimmed as it was between articles
on agricultural tractor design and malnutrition in India—would occupy
the minds of thousands of the world's best scientists and lead to untold
numbers of experiments and tens of thousands of scientific papers to be
published over the next fifty years.

But if the editors at *New Scientist* didn't quite perceive the importance
of the brief article, a sharp-eyed legal minion for Monsanto in Europe
did. Immediately apprehending that Jensen's work was going to create an upheaval, Monsanto's Swedish counsel, Henry Strand, somehow
obtained prepublication copies of the article. Two full weeks before the
issue hit the stands, Strand alerted headquarters in St. Louis to what he
predicted might eventually be a fiasco of global dimensions.

Strand wrote to Monsanto that not only had Jensen found PCBs in
almost all the animal species he analyzed, "Mr. Jensen has also examined the hair of his family and himself and found PCB on all samples.
Most PCB was found in the hair of his wife. But most sensational, was
that the girl aged five months had more PCB in her hair than her brothers and sisters of three and six years. Probably the girl had got the poison
via the mother's milk." The attorney said there was little possibility
that the PCBs Jensen discovered were not Monsanto formulas. He also
added, in what would turn to be the most lawyerly of understatements,
"There is also no doubt that the published facts will cause considerable
unrest in several quarters."

To this bleak news coming from Sweden the Monsanto company
physician in residence responded. Dr. Emmett Kelly, admitting with
surprising honesty that he was "out of his depth" regarding matters of
analytical organic chemistry, still opined that Jensen's painstaking research on PCBs was all wrong. Kelly believed that the peaks on Jensen's

gas chromatograms were most likely produced by herbicides—not Kelly's beloved Aroclor/PCBs. But going public with his theory that Jensen's PCBs were actually herbicide byproducts would have been trouble for Monsanto. "Our only problem," speculated Kelly, "is whether or not we want to bring these facts out and have our herbicide program receive another black eye."

The first patient to be seen with the strange disease was a three-year-old girl with severe acne-like eruptions on her face, in June of 1968. Her parents took her to the outpatient clinic at Kyushu University Hospital, located in the port city of Fukuoka in southern Japan. Within two months, thirteen other people were seen at the same clinic with similar symptoms. All were members of four different households. Local newspapers ran the story in the fall of that year and by then there were more victims—with more symptoms. Along with severe rashes and pustules on their faces, most of the new patients had discharge and swelling around their eyes.

Epidemiologists were brought in to try and pinpoint the cause of the odd illness. They soon determined from questionnaires and interviews that all the victims had consumed the same brand of rice oil, Yusho (meaning "oil" in Japanese), for which they named the mysterious disease. However, even with a name, Yusho was baffling. When tests were performed on the suspect rice oil for causal agents—pesticides, coal tar, machine oil, and the like—no toxic chemicals could be found.

As more victims came to the clinic, the Yusho study team at the hospital brought in a food production engineering professor to work with the manufacturer of the rice oil. Dr. K. Inagami reviewed the lengthy flow chart of the factory and homed in on a heating element that was used to deodorize and decolorize the cooking oil. Questioned as to what substance had been used as the heat transfer fluid inside the element, the plant manager told the professor that it was Kanechlor 400—a Japanese-made PCB under license from Monsanto.

After examining the heat pump filled with the PCBs, Inagami found a minute leak in the device. He then had the remaining unused cooking oil tested for PCBs and all samples were found to be contaminated. To confirm that the illnesses were caused by the specific type of PCB in the rice oil, fatty tissue samples were taken from the victims and analyzed. They too were found to have high levels of the same Kanechlor PCB.

Eighteen hundred patients were eventually treated for Yusho disease in Japan. Probably thousands more consumed the contaminated rice oil but had only subclinical reactions that were either not noticed or not reported. In cases that were reported, the amounts of polychlorinated biphenyl that the victims actually ingested were found to be minuscule: 141 victims were investigated to determine their average intake of PCBs, which turned out to be approximately a single gram—about one twenty-eighth of an ounce.

But even with the tiny ingestion of PCBs by the adult victims, the physical effects were severe and multifarious. In addition to the disfiguring chloracne, problems with eyes and eyesight were the most obvious manifestations of Yusho disease. Upper eyelids of patients were swollen and in some cases, almost shut. There was also a mucous discharge associated with the swelling that never seemed to stop, along with terrible itching—in some cases accompanied by actual loss of vision. The most severely affected adult victims also experienced general numbness in their hands, along with sweating palms and tingling in their feet.

Dermatological problems were also pronounced among the adult Yusho patients. Along with skin eruptions and lesions, there were changes in pigmentation—with the victims' skin often taking on a dark brown color. Small pores became huge blackheads. Their nails often turned black. And even more ominous, many of the adult victims with skin disease also had abnormal liver function test results.

But the worst problems—as would always turn out to be the case with the "magic fluid"—were with childbearing women and infants. Here poisoning with only tiny amounts of polychlorinated biphenyl was lethal. Of the sixteen pregnant women known to have consumed the PCB-contaminated oil, two had stillbirths—an extremely high number statistically. With PCBs clearly able to breach the placental barrier, the babies that survived were often literally marked by their exposure to PCBs in the womb. They had pigmentation changes that left their skin a grayish dark brown—so dark that the Japanese gave them a nickname, calling the Yusho child victims "the black babies."

Obstetricians found that ten of the infants born to Yusho victims had the abnormal pigmentation. And there were other anomalies and birth defects in the Yusho babies. All registered lower birth weights than normal. Some had deformed fingernails and odd gum formations. Luckily over time, the skin pigmentation problems the Yusho children experienced disappeared, but the "magic fluid" would prove to be unforgiving in ways that were far more terrible, as we will see later.

Ten years later, there was an eerily similar disaster in Taiwan that would be called Yu-Cheng disease. Once more, PCBs were found to have contaminated cooking oil. However in this case, Taiwanese and American scientists, already aware of the possible dangers that PCBs posed from the Japanese Yusho calamity, would establish beyond any doubt their lethal toxicity for humans.

As with Yusho, the average intake of PCBs for the adult Yu-Cheng victims was approximately one gram. This minute exposure proved to be fatal for a high percentage of the babies of women who were pregnant when exposed to the PCBs. Of thirty-nine Yu-Cheng babies exposed prenatally, eight died within the first few years of life with respiratory infections being the leading cause of death.

Beyond the high death rates, the researchers found consistent patterns of physical abnormalities in prenatally exposed children that were similar to the deformities found in the PCB-poisoned Japanese infants. Besides chloracne, many were born with odd-looking calcium deposits on their teeth and with skull anomalies. Like their counterparts in Japan, a large number of Yu-Cheng children had hyperpigmentation— extraordinarily dark skin at birth. Many had gums that were enlarged, as well as strangely formed fingernails.

The Yu-Cheng researchers also investigated an array of other effects on the poisoning victims. Animal testing in the years since the Japanese Yusho disaster strongly indicated that PCBs were toxic to humans via multiple pathways and could affect a variety of organs and systems— especially neurologically related ones. To that end, over a ten-year period, the surviving prenatally exposed Yu-Cheng victims were given batteries of psychological and neurological tests.

The results shocked the scientists. Ten percent of the Yu-Cheng children in Taiwan had significant psychomotor damage which was still observable eight years after exposure. The Yu-Cheng children also lagged well behind typical children in their responses to stimuli, and their motor reflexes seemed to be permanently damaged by PCBs. As well, investigators found that even the least affected children had lower IQ scores and suffered from higher rates of attention-deficit-related behavioral disorders. But the worst neurological damage that an unspecified number of prenatally exposed victims of Yu-Cheng disease were found to have was "impaired cognitive abilities." Researchers, trying to save the parents more suffering, labeled many of these child victims as "cognitively impaired" rather than retarded.

(Although there were never thorough tests done on the mental functioning of the fourteen surviving Japanese Yusho disease victims who were exposed prenatally, there were reports of anecdotal evidence gathered by attending physicians that pointed toward significantly lower IQs in the victims—some bordering on mental retardation.)

More subtle findings of this "experiment in nature," as some scientists callously called it, were what PCBs did to the immune systems of victims. Researchers determined that PCBs seemed to target immune functioning by somehow altering critical cellular immune responses in the victims. The Taiwanese children exposed to contaminated rice oil via nursing or direct ingestion had much greater rates of bronchitis and flu attacks when they were infants. Many suffered from continual ear infections up to six years of age; and between the ages of eight and fourteen, Yu-Cheng victims had noticeably higher incidence of middle ear infections. However, the most horrifying findings of the Yu-Cheng investigators involved the reproductive organs of the male Yu-Cheng victims exposed prenatally. Researchers discovered that the boys suffered from high rates of birth defects—abnormalities in their testes and penises, which could only be attributed to exposure to the contaminated cooking oil.

. . . And while the Asian and American scientists documented the dimensions of the Yusho and Yu-Cheng disasters, what were the makers of PCBs doing? They were entangling themselves in perhaps the greatest scientific research fraud in U.S. history.

Industrial Bio-Test Laboratories—IBT—was founded in 1953 in Northbrook, Illinois, by a thirty-five-year-old graduate of the Northwestern University School of Medicine, Dr. Joseph C. Calandra. Within twenty years, IBT would be the preeminent testing laboratory for not only the U.S. government, but for corporate America. The reasons for IBT's success were easy to understand. Dr. Calandra had a reputation for having and maintaining the highest scientific standards and the fees the laboratory charged for testing were reasonable. But most important, Dr. Calandra was known to be an accommodating gentleman who would work diligently to please his corporate clients—especially when they came to him with troublesome products.

So it was natural that since Monsanto had a dozen or so such items that were causing difficulties in the executive suites, they decided to avail themselves of Dr. Calandra's kindness. Initially, after the work of Soren Jensen was made public, Monsanto had done some PCB testing at their own labs. The results were apparently not acceptable, since Monsanto scientists actually confirmed what Dr. Cecil Drinker had found thirty years earlier—that PCBs were highly toxic substances that attacked the liver, causing serious damage to that organ in even low dosages.

To make matters more difficult for Monsanto, a scientist with the U.S. Environmental Protection Agency—Dr. Renate Kimbrough—found that PCBs not only produced benign tumors, but they caused cancer in the rats. Ever wary of the "C" word being attached to their profitable Aroclor line, Monsanto hired an outside consultant to review Dr. Kimbrough's work for mistakes. His conclusion was unequivocal. From his examination, "Dr. Kimbrough's study demonstrated carcinogenicity."

What to do? By 1975, Monsanto decided the best way to defend itself and its product would be to have IBT Labs with Dr. Joseph Calandra take a whack at coming up with some positive research on the "magic fluid." That way they could claim to have obtained competent and unbiased analysis by experts with national stature—not some obscure government scientist, like Dr. Kimbrough.

But even though Dr. Calandra was known for his scientifically obliging and politically gentle ways with big corporate clients, Monsanto apparently wanted to ensure that its payments to IBT were well spent. Thus IBT hired a Monsanto executive and toxicologist to oversee PCB testing along with testing of other Monsanto products that IBT had been contracted to analyze. CEO Calandra even made the Monsanto executive one of his top three associates, putting him in charge of the critical "rat toxicology" department that was doing the most crucial product research for Monsanto.

And so IBT began its two-year-long analysis of the toxicity of Aroclor—the largest research project on the "magic fluid" since commercial manufacturing began in 1931. With the former Monsanto executive presiding over the testing operation, it was not a surprise when the results came in. Contrary to almost all previous animal research experiments over four decades, IBT didn't find that Monsanto's PCBs caused cancer in the test rats. There was some question whether the liver abnormalities that were recorded in the animals were precancerous—but as far as Monsanto was concerned that was just a scientific technicality. Nonetheless, the PCBs were so toxic, causing so much organ damage in the test animals, that even the IBT report writers couldn't bring themselves to say that PCBs were not carcinogenic. Instead they diplomatically decided to designate PCBs as "slightly tumorigenic."

Even this mild conclusion didn't satisfy the Monsanto executives, who'd invested good cash money in IBT to come up with the right findings. In a letter to Dr. Calandra, Monsanto's overseeing executive asked him to change the key line in the research report from "slightly tumorigenic" to "does not appear to be carcinogenic" since the latter phrase was "preferable" for Monsanto. This was not a problem for Dr. Calandra. He promptly made the critical change in the final report for publication and sent in the "before" and "after" versions. Calandra also addressed some disturbing questions that had been raised about "confusion" in the rat data that IBT had provided Monsanto.

• • •

In fact there was a lot more than "confusion" at IBT. Calandra and three other top executives, including the Monsanto ringer, were being investigated by both the FBI and the federal Food and Drug Administration. What the investigation uncovered would result in criminal prosecutions for fraud that would shake the world of scientific research. In court testimony, witnesses told of complete chaos in the "the Swamp"—the name employees gave to the IBT's animal testing laboratory where the Aroclor experiments were done. In two-year-long feeding studies, the animal mortality rates were off the charts, reaching 80 percent. Rats and mice died from drowning, exposure, and cannibalism. Their carcasses decomposed so rapidly that they would ooze through the cages onto the droppings trays. No technician would even enter "the Swamp" without waders to slog through the puddles of water, decaying tissue, and feces.

As bad as things were, it would take an outsider, a veteran Food and Drug Administration pathologist, to raise the alarm. The pathologist— Dr. Adrian Gross—had been reviewing a new round of IBT animal testing data for the government when something caught his eye. A two-year-long study involving the painkiller naproxen (sold as Aleve) looked bogus to Dr. Gross and he immediately suspected IBT of "cooking" the data. "None of the rats had developed cancer," he told an interviewer. "Now any pathologist knows that rats and mice on these long-term studies develop cancer naturally and will have a certain level of mortality. IBT's study said the rats were all clean."

Even after the successful criminal fraud prosecutions of four IBT executives were completed, investigators still couldn't determine the extent of the crime other than to know that it was proportionally staggering. In the decade prior to the Monsanto PCB study, IBT had performed more than 1,500 studies in its main facility in Northbrook. It was estimated that between 35 and 40 percent of all toxicology tests in the United States were performed by IBT laboratories. By any definition, it was a massive compromise of public trust and safety. IBT studies were used to support federal regulatory approval for an array of products: insecticides, herbicides, food additives, chemicals for public water supply treatment, cosmetics, pharmaceuticals, soaps, bleaches, even food colorings used in ice cream. The tests IBT performed—or supposedly performed—on hundreds of consumer goods were used by the federal government to register and approve the products as safe for public use. (As these words are being written, there are apparently still dozens of products being used daily by American consumers that were approved by the government based on IBT tests performed in "the Swamp.")

There is no specific record that the two-year PCB rat studies were faked, but it is almost certain that they were, especially given the exchanges regarding "confusion" about the PCB data coming from "the Swamp." But Monsanto, unlike at least six other large corporate clients of IBT, never sued the company or any of its executives for performing fraudulent studies. Nor did they bother to rerun the trials, as many other IBT clients did. Monsanto took no action, except to pay the legal fees of their shill, a criminal defendant convicted of "fixing" and certifying the results of dozens of fraudulent research experiments.

CHAPTER 5

The Global Poison

*Everyone on earth now eats, drinks, and breathes a constantly
changing and poorly characterized soup of organochlorine contaminants,
most of them unidentified. Moreover, health effects may not appear for
30 years or more after exposure, and the impacts are often seen not
in the exposed individual but in his or her offspring.*

JOE THORNTON, *Pandora's Poison*

Whenever Dr. Wayland Swain reviewed the data in his cramped
cabin, it didn't make sense. Lake Superior was the most unspoiled and
largest body of freshwater in the world, and every fish he and his team
of scientists from the Environmental Protection Agency had caught
and analyzed over the last two years had been contaminated with relatively high levels of PCBs. But it wasn't the amount of contamination
that surprised Dr. Swain, since he was well aware of Jensen's work.
No, what troubled Swain was that all the fish had about the same body
burden of PCBs regardless of where on Lake Superior they had been
caught.

From their ancient converted minesweeper, the *Telson Queen*, Dr.
Swain and his newly formed EPA research group had taken fish from
most of the major river mouths on Lake Superior. They had caught fish
from the middle of the lake, near the surface, and close to the lake's
bottom. All of Swain's results were much the same: PCBs were distrib-

uted evenly and in high amounts in fish throughout the gigantic freshwater ocean.

It was a mystery. There was hardly any industry along Lake Superior
in 1973. The 3,500-mile shoreline of the lake was essentially the same
wilderness it had been for millennia. While there were some factories
built along the larger rivers that flowed into Lake Superior, all of the
fish samples from river mouths showed the same relatively constant
contamination as the fish taken from the most unspoiled areas. So where
could the PCBs be coming from?

Finding out was the mission of Dr. Swain and his crew that summer.
The majority of their work involved sampling new sites on the lake to
make sure that they hadn't missed any major PCB hot spots in previous surveys. On the last day of the cruise, the *Telson Queen* anchored
in Maloney Bay, a harbor for the National Park Service outpost on
lonely, wild Isle Royal. (If you look at a map of Lake Superior with a
bit of imagination, it conjures the image of a wolf's head looking west.
Given this impression, Isle Royal would be the eye of the wolf.) Swain
and his scientific crew had been taking their mid-lake samples from
around Isle Royal for three days, since the waters there were deep and
pure—so pure that when you came inshore you could see a pebble thirty
feet down.

After they anchored on that last day of the cruise, clouds of voracious mosquitoes swarmed toward the old minesweeper, so the crew
abandoned ship and went ashore. After supper they swapped stories with
a Park Service ranger who was stationed on the island. Somehow the
talk turned to Siskiwit Lake, a seven-mile-long lake that was enclosed
within the island.

Filled by pure rainwater, this small lake within a lake was one of
the most pristine, in a scientific sense, in the Northern Hemisphere.
Siskiwit Lake was a deep, oblong basalt bowl with a surface that was
sixty feet higher than Lake Superior, which meant that not even slightly
contaminated lake water could flow into it. It was also untouched by
humans, save for a few hardy campers who would brave the mosquitoes
and make the rough overland trek to its shores.

As twilight fell, one of the *Telson Queen*'s survey crew, thinking out loud, wondered if Siskiwit Lake might make the perfect place to take their control samples of PCBs. Swain immediately agreed. Good background samples—control samples, in scientific parlance—are always critical. Whenever scientists measure any kind of contamination, they need to have samples that come as close as possible to nature at its purest, and Siskiwit Lake seemed to offer the ultimate in control sampling. The researchers could catch untainted wild fish thriving in an aquatic environment without any possibility of known industrial chemical contamination. The next day Swain's team caught their fish samples, ending their survey on an upbeat note.

It would take about six months for all the fish to be analyzed—and it was not an easy or pleasant job, either. In order to measure the PCBs contaminating the fish tissue, each fish has to be measured and weighed, its internal organs dissected and catalogued. Then the fish are pureed into a smelly "smoothie" by putting them into a special Waring blender–like mixer. Fish oils are laboriously extracted from the horrid mash and analyzed with a gas chromatograph fitted with an electron capture detector.

The test results of Dr. Swain's survey on the *Telson Queen* came back around Christmas in 1973. Much to his chagrin, the data seemed to indicate that some manner of laboratory contamination was creating false readings for the EPA's electron capture detector. Swain, a research veteran, knew that lab contamination was not an uncommon confounder of research data, given the strict standards required for organic chemical analyses, so he asked for another round of testing from the government chemists. To his surprise, the EPA lab chief himself sent back the data saying that the results had been recertified and that, in his opinion, there was no chance that they were inaccurate—even if the results were nothing short of bizarre.

The research showed, unequivocally, that the fish taken from Siskiwit Lake, the pristine lake within a lake, had approximately *double* the levels of PCBs in their flesh as the other fish Swain had caught in Lake Superior.

After digesting the startling figures and conferring with his associates, Wayland Swain came to the conclusion that there could only be one possible answer—the PCBs had precipitated out from the air. The contamination was airborne. Siskiwit Lake, with its bowl shape, probably acted like a kind of massive satellite dish, collecting and concentrating airborne PCBs. On a cosmic level, Swain's discovery might have been classified as a huge practical joke by Mother Nature—and a very scary one at that, given the implications. But it was for real. All the time Swain and his team were trying to pin down the source of the toxic PCBs in Lake Superior, they had been breathing it.

It took scientists another ten years to figure out that about 80 percent of the PCBs that were ending up in Lake Superior got there by being transported in the air. And the PCBs contaminating the waters of the rest of the Great Lakes were 60 to 90 percent airborne. In Siskiwit Lake, with no point sources of urban or industrial pollution, atmospheric deposition probably accounted for 100 percent of the load of PCBs.

In basic terms, here's how it happens: When molecules of PCBs precipitate out in the rain, they can accumulate in huge amounts in marine and freshwater systems. After they have built up in an aquatic environment beyond a certain concentration, they start to be exhaled. This odd phenomena, called "outgassing," is caused (in simplified terms once again) when PCB levels in the body of water substantially exceed the amounts in the air over it.

If this happens, the Great Lakes—and any other large body of water—act like gigantic lungs exhaling tons of PCBs right back into the air and into global circulation. Since oceans and seas dominate the surface of Earth, and since PCBs are distributed globally via the atmosphere, the bodies of water on our planet are thoroughly, and relatively evenly, polluted with PCBs, usually in low parts per trillion. (A part per trillion would equal to the width of a hair compared to the circumference of Earth.)

One might conclude that pollution in such infinitesimal amounts doesn't present any environmental threat. Unfortunately, given the ways and means of PCBs, this isn't the case. PCBs undergo a process called biomagnification. When a PCB molecule leaves the atmosphere in a raindrop and reaches the surface of the earth, it frequently becomes part of the microscopic biological community that is nearly everywhere in the biosphere. A simple story line for the biomagnification process might go like this: Tiny aquatic invertebrates absorb the PCB molecules as they feed in water contaminated with parts per trillion of PCBs. Since PCBs are lipophilic and gather in fatty tissue, these microscopic creatures begin the biomagnification process by concentrating PCBs in their fat at hundreds of times the amounts to which they were exposed in their watery habitat.

In turn, the contaminated invertebrates are food for midget aquatic carnivores such as krill shrimp or perhaps smaller-sized schooling fish. Again the PCBs are greatly magnified in the tissue of these animals— this time to parts per billion. (A part per billion would be one second out of thirty-two years.) Then even larger predators—flesh-eating birds, salmon, bass, and trout, for example—ingest contaminated prey and absorb their own magnified doses of PCBs. The tissue of these creatures is contaminated in parts per million. (A part per million would be one square inch of a baseball field.) These predators in turn are consumed by the highly evolved, warm-blooded mammalian carnivores such as mink, seals, bear, otter, whales, and human beings—further concentrating the PCBs.

The result of the biomagnification process is that PCBs, and a variety of other industrial chemicals, can be magnified exponentially in the bodies of high-level predators—like us.

All PCBs circulating in the atmosphere or in the world's aquatic environments began their geophysical lives on land, more than half of them manufactured in just two Monsanto chemical plants in the United States—at Sauget, Illinois, and the original Anniston, Alabama, plant built by Theodore Swann. Understanding how PCBs are transported is a hard scientific problem, not only because of the supercomplex dynamics of Earth's biosphere, but also because there have been 209 different varieties, or congeners, of PCBs manufactured. Each variety has its own unique molecular structure, with subtle but often quite different toxic effects, depending on nothing more than the position of a single atom within the molecule.

The mysteries of PCB behavior are compounded further because all 209 congeners degrade differently in the environment, depending on the medium in which they are found. A "light" variety such as PCB-118 could morph into a dissimilar substance in fatty tissue or blood, whereas the heavier PCB-153 might keep its chemical character. And every variety of PCB can be partially transformed into metabolites, or by-products, by living organisms.

Biologically speaking, some PCBs act like supertoxic substances such as furans and dioxins. Some don't. Some PCBs block production of specific hormones, while others—with almost identical molecular structures—will promote the release of those same hormones.

All PCBs contain some amount of chlorine. While there are exceptions, the heavier PCBs—those with a higher percentage of chlorine atoms—tend to degrade more slowly than lighter, more volatile PCBs. According to a simple scientific concept with an overly complicated name—the global fractionation theory—PCBs migrate from warmer climates to colder ones. Because they volatilize more readily when heated and since heat is greater on Earth's surface in the lower latitudes, PCBs tend to eventually come to rest in the coldest regions of the globe—the Arctic and Antarctic.

The geophysical behavior of PCBs produces bizarre and counterintuitive results that only substantiate the mysterious and fickle ways of the "magic fluid." One of the worst cases involves Inuit Native Americans, who live at the world's most northern latitudes. These native people—living as far from sources of pollution as anyone in the Northern Hemisphere—are the most PCB-contaminated human beings on the planet.

However the most surprising and unforeseen type of contamination by PCBs is called biotransportation. Researchers studying remote lakes near the Arctic Circle in Alaska discovered that the bodies of salmon that had migrated and died in the lake after spawning were so contaminated with PCBs that grayling—a nonmigrating species of fish—had twice as much PCB in their tissues as they did when living in lakes where contaminated salmon didn't spawn and die.

In any case, predicting how quickly any particular variety of PCB will break down in the biosphere is like a guessing game without set rules. According to laboratory experiments, the half-life of "average" PCB molecules may be greater than twenty years in the atmosphere. In water and soil, their half-life might be closer to forty years. And in places such as deep ocean trenches and old landfills where there may be little sunlight or bacterial activity, the heavier, more chlorinated PCBs may remain for millennia, perhaps many millennia.

According to the United Nations, approximately 1.2 million metric tons of PCBs were produced between 1929 and 1977. An undetermined amount was manufactured after that in Eastern Europe, the Soviet Union, and the People's Republic of China. Since it is not in the best political or public relations interests of either governments or corporations to release accurate PCB production figures—especially if they are high—one must conclude that thousands, perhaps tens of thousands of tons, were manufactured but never acknowledged.

And where are all of those PCBs? Again, there are only gross estimates available, but the best guess by scientists is that only about one-third of all the PCBs produced are now circulating in the environment. Approximately 4 percent have been destroyed by intentional human intervention. The other approximately 60 percent of all the PCBs manufactured are still out there, mostly in so-called "closed" systems such as hydraulic controls, capacitors, and transformers.

We are all contaminated. All humanity is part of a global, ongoing lab experiment without walls—an unsanctioned one, to be sure. And regardless of any controversy that surrounds the question of PCB toxicity, the fact that all of humanity has been contaminated is incontrovertible.

Most humans have a mean average of approximately one part per million of PCBs somewhere in their tissues, and approximately one to ten parts per billion circulating in their blood. These mean averages, as might be expected, are quite variable. People in the Northern Hemisphere have been found to have higher body burdens of PCBs than their southern neighbors. With the exception of infants who have been breastfed, older people usually carry more PCBs in their tissues than do younger ones, since PCBs can accumulate in the body with age.

Metabolism also plays a role; the more lipid tissue you have (the fatter you are) the more PCBs you may have since PCBs have such a strong affinity for fatty tissue.

In addition, genes may play an important, but not well understood, role in how we each carry and react to our body burden of PCBs. For example, some very elderly people test with extremely high levels of PCBs in blood serum and apparently have retained those levels for decades without obvious ill effects. Does this indicate that keeping PCBs suspended in the blood is a healthier natural coping method than having them locked in tissues? And is the ability to suspend PCBs in the blood a genetic trait or perhaps some function of the type of PCBs that an individual was exposed to?

Unfortunately scientists don't know.

However, what is known through blood serum and lipid tissue studies is this: the human race all over the planet is contaminated with PCBs (along with dozens of other industrial chemicals). The following smattering of evidence from international studies gives some idea of the pervasiveness of human PCB contamination:

- In Russia, eighteen- to twenty-four-year-old women had mean blood serum concentrations of PCBs that were in the seven parts per billion range. Older Russian women had double that amount.
- In Croatia, mean average levels for adults were 160 parts per billion for certain "marker" PCBs.
- Roman women were found to have 900 parts per trillion PCBs (almost one part per billion) in the deepest recesses of their ovaries.
- In New Zealand, where the government undertook the only nation-wide study of environmental contaminants yet completed, the citizens tested who were over fourteen years old had mean averages of seventy-nine parts per billion of PCBs in their blood.
- Samples taken from abdominal fat of Inuit in Greenland in a mid-1990s' study revealed a mean-average of more than fifteen

parts per million. (This level of contamination made the Inuits not only the most contaminated people on the planet as far as PCBs were concerned, but placed them thirteenth on a list of the most PCBs-contaminated species on Earth.)

In the United States eleven studies of more than four thousand subjects found mean levels of PCBs in blood serum ranging between three and fifteen parts per billion. One of the most provocative and interesting research projects involving PCB contamination in Americans was completed in 2003. The Environmental Working Group, a Washington-based think tank, had the blood of nine volunteers tested for the most common industrial toxins. All of the volunteers (one of whom was PBS's Bill Moyers) were well-heeled intellectuals—purposely not a part of the population that would have work-related exposures.

Even with these elitist parameters, all nine were found to be infested with some forty-eight different varieties of PCBs. The PCB levels found in the EWG blood serum study were broken down by congener—not by the total amount of PCBs. Consequently the levels were usually in the parts-per-trillion range. However, PCBs—even in those minute quantities—made the "magic fluid" the leading industrial chemical contaminating all the EWG volunteers. And as the study pointedly related, the levels were measured almost thirty years *after* PCBs were banned in the United States.

Besides human blood, breast milk has been the focus of many PCBs studies because of its high fat content. But a definitive picture of the problem of breast milk contamination by PCBs is still not at hand. Research comparisons of human breast milk across ethnic groups and nations are often apples-and-oranges affairs due to changing and differing analytical techniques for measuring PCBs. Nevertheless, human breast milk—with the possible exception of wild orcas' milk—turns out to be, almost surely, the most PCB-contaminated, naturally produced fluid on the planet.

PCBs average between one-fifth of a part per million to almost one part per million in breast milk globally, but the levels vary by geography. Ireland and Bulgaria had the lowest levels and the highest were found in the Czech Republic. Norway, Sweden and the Ukraine, along with the United States were somewhere in the mid-range of breast milk contamination documented in international research investigations.

Whatever the body burdens of PCBs that we all carry, most comes from the food we eat. Since PCBs are lipophilic and tend to congregate in fats, the foods highest in PCB concentrations are dairy products and cold-water, fatty fish. In the United States, the Food and Drug Administration has set limits on PCBs in food from a maximum of 200 parts per billion in infant foods to a maximum of three parts per million for poultry (though as far as I know, the FDA conducts no regular testing for PCBs in these foods). And as we will see, these limits do not lessen the dangers that PCBs present to humans.

So PCBs contaminate the entire planet, but what are the "temporal trends" (as the scientists would say) of PCBs? Rephrasing the question in everyday terms: Are PCBs being detected in increasing amounts? Decreasing amounts? Or are their levels staying the same? The simple answer for human beings is that, since their broad banishment in the 1970s, PCBs are being generally detected in lessening amounts in milk, fatty tissue, and blood in many areas—but not all.

From 1993 to 2002, the World Health Organization (WHO) found that PCBs declined in human milk by about 40 percent. In lipid tissues taken mostly from cadavers, PCBs have been declining in Europeans at a rate of approximately 5 percent per year since the early 1990s, again according to the WHO. In northern areas of the globe, where PCB-contamination of native peoples such as the Inuit is extremely high, there has also been a decline in breast milk contamination although less marked. As an example, in one of the most common and toxic PCB congeners, PCB-153, the decline in Inuit mother's milk was from a high of 220 parts per billion in the early 1970s about 150 parts per billion in the early 1990s.

Unfortunately the downturn of PCB body burdens in humans doesn't mean that the global threat to health that PCBs present is over. Far from it. Cancers related to immune disease, from mild to catastrophic, shown to be associated with body burdens of PCBs are on the rise. And counterintuitive as it may sound, the reduction of PCB body burdens could theoretically *increase* its toxicity under certain circumstances for humans, which we will find out about a bit later.

But the downward trend seen in humans is not nearly so pronounced for the rest of Earth's biota. PCBs are declining in some areas, staying the same in others and actually increasing in a few scientifically important regions such as the mid-Pacific ocean. In Japan, a country that has carefully monitored PCBs after the Yusho disaster, researchers found that fish taken from the bays of Tokyo and Osaka had seen average levels of PCBs decrease in the 1970s, but found that there were no decreases of PCBs in fish since then. Lake Baikal, the oldest freshwater system of its size on Earth, and an area intensively studied by Russian scientists, has shown no decline in PCBs levels through the 1990s and its populations of seals remain at risk of extinction because of reproductive failures that researchers believe stem, at least partially, from PCBs and PCB-cohorts.

While cases of increasing loads of PCBs are relatively rare, the exceptions are vexing and even menacing. In ivory gulls—a migratory marine bird of the North Atlantic—scientists recently documented a 50 percent increase in PCBs in their eggs since the early 1980s. Even more ominous is a study done in 2006, where investigators measured PCB levels in Pacific black-footed albatrosses and found significant increases in PCB body burdens in the birds. What is most worrisome about this new data is that the black-footed albatross is a top-of-the-food-chain predator that feeds in the vast central Pacific region where there are no industrial sources of pollution.

Still, most temporal trends for PCBs can be described as a "ski-jump" curve on a graph—sharply downward in the late 1970s and 80s—and leveling off in the 1990s to a straight line or sometimes a slightly upward trend. But even after being banned decades ago in most industrial nations, PCBs are not showing any indication that they will disappear from Earth's environment completely even if the downward trend continues—which it may not, since only one third of all the PCBs manufactured have reached the biosphere thus far.

Given this reality, perhaps the most germane question becomes this: if the "magic fluid" isn't going to go completely away for perhaps millennia, do the levels that remain constitute a serious threat to us and the rest of Earth's biota? The answer, as of the last ten years worth of research by some of our best researchers, is yes, and in ways we could never have imagined.

CHAPTER 6

PCBs and Kids

Is it enough to have one long-term study of humans that shows negative effects,
even if the results of other studies do not completely agree?
Is it enough to have laboratory evidence that a pollutant disrupts
brain chemicals and hormones in developing animals?

COLLEEN MOORE, *Silent Scourge*

The poisoning of thousands of people in Japan and Taiwan from eating PCB-contaminated rice cooking oil produced databases that conclusively determined that PCBs were extremely toxic to human beings, even in minuscule amounts; however, most of that research involved the gross effects of PCBs—infant mortality, skin diseases, eyesight. But the most tragic aspect uncovered by investigators was not the easily observable effects, terrible as they were. The worst damage was to the most vulnerable of human beings—infants and young children.

This part of our story begins with an Asian scientist doing something that most Western researchers—to the detriment of science—would not do: he measured the penis size of the surviving Yu-Cheng boys—the children exposed prenatally to PCBs when their mothers ingested contaminated cooking oil. Those tests revealed that the victims were found to have consistently smaller than normal penises. Aside from the titters that this result always seems to elicit from those who are less than empathetic, the finding was highly significant because it confirmed a key toxic trait of PCBs—the ability to affect male reproductive organs.

Nevertheless, the defenders of PCBs in the scientific community could point out that the Yu-Cheng women had eaten highly contaminated rice oil (although the actual amounts ingested were very small) and their PCB exposure rates were thousands of times greater than what a pregnant women would receive from a normal diet, whether Asian or Western. But PCBs are tragically confounding. As it would turn out, some mothers didn't need high-dose exposure from PCBs to have far more terrible consequences for their sons than small penises.

In the United States, according to Dr. Jane Houlihan of the Environmental Working Group, "the incidence of testicular cancer rose 41.5 percent between 1973 and 1996, an average of 1.8 percent per year." This made it the most common cancer for men between the ages of fifteen and thirty-five. In Great Britain, the occurrence of testicular cancer has doubled since 1975. In 2001, almost 1 percent of Danish men were being treated for testicular cancer.

When epidemiologists see a large rise of certain types of cancer over short periods of time, they usually suspect that the cause is environmental. In other words, some substance—or combination of substances that the victims have come in contact with—is causing the cancer. But in any investigation of cancer etiology, epidemiologists also realize that there may be a genetic component that allows the victim to be more susceptible than others to the causative agent, although the cancer might be unlikely to occur without an environmental trigger.

Ever since Soren Jensen's discovery that PCBs were everywhere in the biosphere, Swedish scientists have been in the forefront of research into the effects of chemical contamination. Therefore it was not all that unexpected when, in the winter of 2002, a Swedish scientist executing a groundbreaking research regimen discovered what could be called a virtual "smoking gun" linking PCBs with testicular cancer.

But, as always, nothing turns out to be simple with Ted Swann's "magic fluid." The association found between testicular cancer and PCBs was powerful but circuitous. Swedish oncologist Dr. Lennart

Hardell recruited three dozen men diagnosed with testicular cancer, along with a like number of controls (cancer-free men of similar background) for his study—sixty-one subjects in all. Neither the controls nor the testicular cancer victims were known to have ever been exposed to high levels of PCBs through their workplace or their home environment.

Dr. Hardell carefully matched the control group and the victims of testicular cancer by age, keeping them within five years of each other. The median age was thirty years old. He then tested all the subjects—for both PCBs and other toxic industrial chemical cousins such as DDE, a DDT metabolite. Hardell found no significant difference between the cancer victims and the controls as to PCB levels in their circulatory systems. The same held for the other toxic compounds: no differences between the victims of testicular cancer and the cancer-free controls.

Then Hardell did something that hadn't been done before in researching PCBs and their relationship to cancer. He began testing the mothers of those in the study group. What Hardell discovered will probably alter the course of reproductive system cancer research. The test results showed a strong link between the amounts of polychlorinated biphenyl and, to a lesser extent, other industrial chemicals in the mothers' blood and the incidence of testicular cancer in their sons. The link was so robust that statistically it would be improbable for it to be the product of methodological error on the part of Dr. Hardell and his colleagues, even with the moderate size of the test group.

What made the Hardell research so significant was the fact that there were no known poisoning victims in the study group—no Yusho or Yu-Cheng mothers—just normal women, eating normal Northern European diets. In fact, the mothers of the cancer victims had levels of PCBs in their blood that were not greatly above the background PCB levels that anyone in an industrialized country might have.

The study results begged an obvious question of Dr. Hardell: just how could the PCBs cause testicular cancer in the sons, but not cause some type of cancer in the mothers? The answer would take some detective work. The average study victim was born in the early 1970s, a time when their mothers would be exposed to the highest levels of PCBs, be-

fore all varieties were banned in Sweden. Scientists estimate that certain
PCBs can stay in blood serum anywhere from seven to thirty years after
exposure. Thus, the differences in blood levels between the mothers of
the cancer victims and the control mothers was probably a reflection
of differences in their diets—with the victims' mothers perhaps eating
somewhat more Baltic fish, well known for having highly elevated levels
of PCBs contamination.

Both the controls and the testicular cancer victims would have been
exposed to high infusions of PCBs during breast-feeding as infants.
(Dr. Hardell found that most of the women in the study breast-fed their
sons, and did so for reasonably similar lengths of time.) But even though
the controls also got a big burst of PCBs during breast-feeding, they
didn't develop testicular cancer. This led Dr. Hardell to conclude that
the PCBs were doing their carcinogenic damage prenatally, especially
since PCBs were understood to cross the placental barrier, as had been
demonstrated in the Yusho and Yu-Cheng disasters.

Prenatal exposure could clarify why the mothers didn't get cancer
while their sons did. The mechanism might be explained this way: while
parts per billion of PCBs in blood serum might not cause cancer in an
adult woman, those same levels of PCBs might be carcinogenic for a
developing male fetus weighing only two ounces—some one thousand
times smaller than a 120-pound adult female.

Furthermore, the "magic fluid" was already known to have insidious
hormonal effects that could be the prenatal causative agent in the testicular cancer epidemic. PCBs are generally the most ubiquitous chemical
of a group of toxic synthetic compounds called "endocrine disrupters." These synthetic industrial contaminants can—and do—mimic
powerful hormones such as androgen and estrogen, which play critical roles in the formation and development of reproductive organs in
mammals.

The male reproductive maladies thought to be caused by endocrine-
disrupting chemicals such as PCBs are grouped under a single heading
called Testicular Dysgenesis Syndrome, according to a theory developed by Danish researcher Dr. Nils Skakkebaek and his associates.
Along with testicular cancer, Testicular Dysgenesis Syndrome includes

undescended testes, genital tract abnormalities, and poor sperm quality. All are understood to be related diseases and therefore are grouped together under the same syndrome heading.

Essentially, Testicular Dysgenesis Syndrome is believed to be triggered in the male fetus by abnormal estrogen exposure in the womb. What happens, in abridged form, is that through estrogenic activity, the production or correct replication of male reproductive cells—called Sertoli cells—is either inhibited or somehow disrupted. These Sertoli cells have lots of responsibilities. How many Sertoli cells a man has will determine how much sperm he is able to produce, so fewer Sertoli cells means less sperm output. The Sertoli cells also regulate the descent of the testes and the formation of the reproductive tract, including the placement of the urethra.

It is within the developing testes that the Sertoli cells regulate cell division. And it is here that vanguard scientists like Dr. Skakkebaek believe that abnormalities take place; some sort of cellular processing error occurs that destroys or damages cells. These damaged cells—the processing mistakes—are what may give rise to testicular cancer after puberty. (However, what is completely unresolved is just how these damaged cells are organized during fetal development and what causes their proliferation into full-blown testicular cancer, often decades later.)

Along with possible environmental factors such as PCBs and other industrial chemical contaminants, most scientists suppose that genetic tendencies play a role. It may be that people who belong to certain ethnic groups or have similar genetic backgrounds are less able to repair cellular damage from PCBs. Or, to make things even more complicated, they may be able to resist damage from one variety of PCBs and not another.

Regardless of the obstacles in the path toward complete understanding of PCBs' potential for causing testicular cancer in humans, literally dozens of animal and in vitro laboratory studies have implicated PCBs as a causative agent for reproductive abnormalities in a variety of test species from rats to chickens. Additionally, feed contaminated with only low parts per billion of PCBs has been proven to cause total reproductive failure in some mammals such as mink. But as yet there have been no laboratory studies where PCBs have been shown to cause testicular cancer in any test mammals. So once again with the "magic fluid," one mystery enshrouds another, and another and another . . .

The late Dr. Wayland Swain, the man who discovered that PCBs were contaminating the planet by being wafted on currents of air, also was the catalyst for the discovery that PCBs were damaging the brains of children. How he intuitively shifted his concerns about PCBs from the ecological condition of the Great Lakes to the condition of human children, we don't know. But he did, and it started another investigative/scientific paradigm.

In the late 1970s, Dr. Swain convinced a bright, young husband–wife team of child development researchers, Sandra and Joseph Jacobson, to investigate whether the amounts of polychlorinated biphenyl in a "normal" diet could be harmful to children neurologically, offering financial support for the initial research program. The Jacobsons, freshly finished with their PhD studies at Harvard (where Joseph also received a law degree) and full of youthful vim, leapt at the chance. It would be a fateful decision for them and for the world's understanding of PCBs.

To say the least, the Jacobsons were facing an intimidating task. Nobody had ever attempted to analyze how PCBs—or any other industrial chemical, for that matter—could affect the minds of humans. There was strong evidence from the Yusho and Yu-Cheng cooking oil disasters that there was a serious neurological impact on both adults and children from the ingestion and exposure to PCBs, but that data came from a population that was poisoned en masse.

There was nothing concrete linking PCBs with neurological damage in populations with normal regional diets because human toxicity studies are difficult to formulate and even harder to execute, given the intricate interfaces between biology, chemistry, and sociology. Nor, of course, can potentially harmful experiments be ethically performed on humans—and studies with toxic substances such as PCBs pose known dangers.

So the Jacobsons were hard up against a high wall. They would have to come up with a study model that would isolate PCBs in the diet of a human population and then make comparisons with control groups. These were tough tasks to be sure, first, because PCBs are omnipresent, which means that a scientist simply cannot find any control subjects that don't have some PCBs somewhere in their bodies—even if the amounts aren't detectable by most analytical methods. And secondly, no one would knowingly eat a diet high in PCBs, since their toxic reputation had already gotten wide media coverage by the late 1970s.

With the aid of the ever-helpful Dr. Swain, the Jacobsons came up with a solution that was both dazzling scientifically and numbing logistically: in 1980 and 1981 they coordinated more than eight thousand written interviews with women from four western Michigan hospitals who had delivered babies. Among these mothers, the Jacobsons found 240 who had eaten more than twenty-six pounds of Lake Michigan fish in the previous six years. After identifying the women with high-fish diets, the Jacobsons randomly picked a control group from women who reported that they hadn't eaten any Lake Michigan fish in the six years prior to their pregnancy.

Why fish eaters? Because Lake Michigan fish—lake trout and salmon—were fatty. And fatty fish, especially in colder northern climates, had comparatively high levels of lipid-loving PCBs. Why pregnant women who had delivered babies? Because that way the Jacobsons—using the controls who hadn't eaten Lake Michigan fish for comparison—could determine if there was a relationship between dietary PCB intake of the mothers and neurological functioning in their children. To make matters even more complicated scientifically, the Jacobsons planned to do their study longitudinally, meaning they would follow each one of their child subjects until adolescence—more than ten years.

The Jacobsons dove in. All the mothers were tested for exposure to PCBs by measuring their blood serum levels and breast milk levels, and in an interesting twist, the PCB levels of their infants' umbilical cords were analyzed. Along with chemical analyses, social and economic status was noted for all the mothers.

For the first part of the neurological study, newborns were tested for a variety of reflexes such as sucking (when an object touches the lips or mouth), rooting (when an object touches the infant's cheek and the baby turns in that direction), the grasping reflex (when an object touches the infant's palm and the infant grasps it firmly), and the Moro reflex (which measures how a baby reacts to being startled).

The Jacobsons' earliest test results were not definitive (nothing would ever be with PCBs), but they were illuminating. About one third of the newborn PCB-exposed infants showed lower than normal reflexes—a "worrisome" indicator of less than robust neurological health. The PCB-exposed infants also showed lower levels of motor development, and during testing they were not as active as the control infants whose mothers hadn't eaten contaminated lake fish.

At seven months old, the infants were tested again, this time using the widely accepted Fagan test for infant intelligence, a type of early memory assessment that predicts verbal IQ. Here babies are shown two pictures that are identical and then the researcher replaces one of the pictures with a new one. Normal six-month-old infants are known to have a visual preference for the most novel item in their view, watching and even grasping at it. However, the children with the highest exposures of PCBs didn't appear to recognize that one of the pictures had been replaced.

All the children—a total of more than three hundred in the study—were tested again at four years old. This new battery of tests was extremely thorough, with a combination of standard intelligence tests, lab exams that assessed memory, and tests of attention and the ability to make quick visual discriminations during tasking. The children were physically measured and their growth rates were recorded, as well.

Again PCBs were linked to mental diminution. The children who had the highest exposure levels performed substantially more poorly on memory tests. They also tended to be less able to complete visual tasks quickly and were usually slower in their overall thought processing. Prenatal exposure to PCBs also appeared to have an effect on the children physically. The more exposed children were somewhat smaller and less energetic than their control counterparts.

(Interestingly, some 7 percent of all the children in the study were unable to complete the test regimen because of apparently untoward behavior with the researchers who were trying to take measurements. The Jacobsons provided no information regarding the PCB exposure rates of these children—or whether they belonged to the exposed or the control group. This may have been an important omission, since PCBs were known to be strongly correlated to difficult behavior—attention deficit–type disorders—in both the Yusho and Yu-Cheng children and are now suspected to be a factor in the etiology of autism, as we'll see later.)

As far as the actual mechanisms that caused possible brain damage via exposure to PCBs, such speculation was outside the scope of the Jacobsons' work. Nonetheless there is substantial evidence that one avenue of neural toxicity for children, especially prenatally, is through disruption of thyroid functioning. According to Dr. David O. Carpenter, a researcher at the State University of New York at Albany—and a globally respected expert on toxicology—PCBs may do their damage by altering the levels of thyroid hormones in the blood during critical periods of brain development in children, and especially in fetuses. Carpenter speculated that the chemical structure of some varieties of PCBs is so similar to that of thyroxine—a key thyroid hormone—that the likeness could possibly cause the brain to misinterpret the PCBs as hormonal signals from the thyroid that regulate normal brain growth and functioning.

Another line of research on the question of how PCBs damage the neurological systems of children posits that PCBs interfere catastrophically with the brain's ability to form complex nerve networks. Specifically, Dongren Yang of Oregon Health and Science University, together with his collaborators, discovered that the most commonly found PCB (Aroclor 1254) compromised the ability of rat pups to form normal elongated dendrites, crucial nerve fibers that allow mammals to store and recall experiential memories. In other words, rat pups exposed to PCBs during gestation and lactation were "dumbed down"—unable to correctly perform maze tests that the unexposed rats could navigate with relative ease. (Most interestingly and counterintuitively, the rat pups dosed with the smaller amounts of polychlorinated biphenyl suffered the worst damage to their brain cells. Yang's experiments were another confirmation of the theory of "low-dose" effects, meaning that smaller doses of certain toxins like PCBs can actually be *more* damaging than larger ones. We will examine this important phenomenon in later chapters.)

Every parent knows that children can outgrow problems of all sorts— both mental and physical. So perhaps the premier question about PCBs and their impact on the neurological health of children was: could PCB damage be permanent? The unfortunate answer, obvious from interpretations of the Jacobsons' longitudinal research, was yes.

All the children in the study were retested again at eleven years old. After a careful assessment of the data factoring in social and family differences, the more highly exposed children still carried cognitive deficits from their PCB exposure. Their IQ scores were lower by an average of approximately 5 percent, with the verbal segments of their tests showing the most deficiency. The eleven-year-olds with high PCB exposure also fared comparatively worse in reading comprehension and were more easily distracted, leading to concerns about threshold hyperactivity disorder.

However, the Jacobsons found no relationship between the cognitive deficits and *postnatal* PCB exposure. Even though infants might have received significant doses of PCBs through breast-feeding, the exposure didn't appear to affect their test performance. This led the Jacobsons to surmise, as Hardell did two decades later, that the neurological damage PCBs were doing was caused by fetal exposure.

Still there wasn't the smoking gun. Of their findings the Jacobsons wrote, "There was no evidence of gross intellectual impairment among the children we studied." But, they continued "there was a substantial increase in the proportion of children at the lower end of the normal range" and they "would be expected to function more poorly in school. This intellectual deficit seemed to interfere particularly with reading mastery." The *New England Journal of Medicine* published the final results of the Jacobsons' comprehensive and meticulous PCB research in 1996, after the data on eleven-year-olds had been fully evaluated.

Even though the Jacobsons were too cautious to draw a direct line between PCBs and neurological damage in children, their innovative study nevertheless drew stern criticism and disparagement from General Electric, which was then under federal government scrutiny and citizen activist pressure for blatant and chronic contamination of the Hudson River and other Superfund sites with PCBs. The GE assault on the Jacobsons' Michigan study was a standard assault of its kind, picking aspects of study design that could be questioned, with the knowledge that no scientific research involving human biology could be flawlessly conceived and executed, let alone one with the scope of the Jacobsons'.

Scientists paid by GE or industry front groups sniped at perhaps the most vulnerable dimension of the study, and of most epidemiological studies: the legitimacy of the control group. For their controls, the Jacobsons had randomly picked mothers who didn't have a history of eating PCB-laden sport fish, from the same Michigan hospitals where the high-exposure mothers had delivered their babies. It turned out that these randomly selected control mothers had backgrounds that were somewhat dissimilar to those of the high-exposure mothers in ways that might conceivably distort some of the results.

The control mothers were on average significantly heavier at delivery time. They were younger and less likely to have consumed alcohol during their pregnancy and were half as likely to be coffee drinkers. All these behaviors would weight the control mothers toward having children who would be, on average, healthier—both neurologically and physically. Furthermore, the Jacobsons didn't take into account the potentially considerable impact of secondhand smoke—that is, exposure of the pregnant mother to a smoker or smokers in their home even if they themselves didn't use tobacco. (In defense of the Jacobsons' study design, the dangers of secondhand smoke had not been known at the time that their work began.)

And finally, the General Electric critics used the effectively impregnable position of all apologists for the "magic fluid." Since PCBs were just one of dozens of industrial chemical contaminants that essentially all human beings carried within them, and since the Jacobsons tested only for PCBs, how did they know that it wasn't another chemical contaminant in the mothers' system that was causing the neurological damage? *If*, that is, there even was neurological damage—something the GE scientists wouldn't even fully concede.

Defenders of the Jacobsons' work could cite literally dozens, if not hundreds, of animal research studies, both of primates and nonprimates, where PCBs were specifically isolated in the subject animals' diets and found to have caused substantial neurological damage. Across species, PCBs significantly reduced body weights of the test animals at birth. Prenatal exposure to PCBs was also shown, time and again, to cause behavioral changes in monkeys, mice, and rats. Their ability to learn, as well as other types of higher cognitive functioning, was degraded by exposure to PCBs in minute quantities—quantities similar to what human beings would be exposed to in a normal diet.

However the primary criticism of the Jacobsons' research—that their studies of PCBs had serious design flaws that nullified their Herculean effort—would be defeated, not by the Jacobsons themselves, but by a broad and powerful coalition of researchers in the United States, Europe, and Japan who were inspired by the work of the husband-and-wife team.

In the United States there were major comprehensive follow-up studies of the neurological impact of PCBs on children done in North Carolina, New York (four studies), Massachusetts, and Wisconsin (three studies). In Canada, there was a study of the effects of PCBs on Inuit children done in the Hudson Bay region. In Europe there were multiple studies in Sweden, as well as two in the Netherlands, and more in Denmark, Germany, Croatia, and Slovenia. There were also three studies in Japan, not including the original Yusho and Yu-Cheng research. These studies validated the Jacobsons' findings time and again, with more than 90 percent finding significant associations between PCBs and serious neurological problems in children exposed to them in even modestly elevated levels.

Research not involving cognitive effects found negative correlations between PCBs and many different aspects of children's health. Ten studies linked PCBs to lower birth weights. Five found an association between PCB exposure and shorter pregnancies. Two studies found that newborns exposed to higher levels of PCBs had smaller head sizes. A half dozen research programs that analyzed the impact of PCBs on the physical growth of children all found that the PCBs were associated with impaired growth in babies and toddlers. Eight of nine studies that investigated the relationship between PCB exposure and immune system response in children reported impairment.

Of the immune/PCB studies, most showed a disturbing reduction in cellular immune responses. Children with high exposures to PCBs had a significantly increased incidence of respiratory infections and many had "decreased rate of take of vaccinations"—both strong indicators of damaged immune systems. The most common finding by researchers examining the relationship between elevated exposure to PCBs and immune system impairment in children was severe ear infections. In the Inuit study in Canada, more than 80 percent of the subject children had experienced at least one occurrence of acute ear infection. (Since these episodes occurred regardless of whether the children were breast-fed or bottle fed, the insults by PCBs to the children's immune systems evidently occurred prenatally.)

So PCBs, as far as infants and children were concerned, were bad actors—very bad actors—with the weight of evidence being overwhelming. However, none of the studies could absolutely point to PCBs as a causative agent. There were just too many scientific subtleties, too many variables, too many other industrial chemicals that were contaminating all the children.

Nearly as difficult as determining if PCBs and/or their metabolites are causing disease in humans is the task of pinpointing the cause of fluctuations in the sex ratios of human populations that researchers have noted with alarm, starting in the late 1980s. Since the Asian PCB tragedies, a number of studies have looked at this critical scientific conundrum because skewed sex ratios can mean the eventual, inexorable extinction of a species if not halted. Some of the studies showed a significant decline in male births to women exposed to PCBs. Other studies found that male births actually increased in the populations studied.

However, what seemed clear to investigators was that PCBs and their metabolites, such as dioxins, were having some sort of impact on the ratio of baby boys to baby girls.

Scandinavian scientists—led by Henrik Møller—noted a significant decline in male births in four countries: Denmark, Finland, Norway, and Sweden. Searching records back through 1945, Møller and his colleagues determined that the declining trend in births of male babies was remarkably similar among the four countries. Møller also concluded—even more ominously—that "decreasing sex-ratios in populations may be part of a scenario of increasing male reproductive hazards, possibly also comprising testicular cancer, male genital malformations and declining sperm quality. The specific agents that may have caused the observed trends remain to be identified, but agents that act prenatally to disrupt normal development and differentiation of the male reproductive organs may be particularly relevant." Halfway around the world, Japanese researchers, in their meticulous way, found that PCBs hadn't skewed the sex ratios of the children born to women who were victims of the infamous Yusho poisoning, but they did determine that men exposed to PCBs in the incident were far less likely to father a male child.

The most stunning research on PCBs and skewed sex ratios came from two investigators at the University of Aarhus in Denmark. Lars-Otto Reiersen, head of the Arctic Monitoring and Assessment Programme, and Jens Hansen, director of the Centre of Arctic Environmental Medicine discovered that across the entire arctic and subarctic regions almost twice as many girls were being born as boys. Reiersen said, "When the mother had an average of two to four micrograms of PCBs or more per litre of blood, we found she bore on average two girls for every boy." In some areas, the skewing is apparently catastrophic. In Greenland, government officials report that near the U.S. air base in Thule (an area thought to be highly contaminated by PCBs from years of military operations) "only girl babies are being born to Inuit families." Unfortunately poor nutrition and alcoholism are serious problems in the arctic and these issues tend to confound air-tight research results, so the scientists could only speculate on PCBs causing the disaster.

However, in July of 2008, the smoking gun linking PCBs to sex ratio was finally found. A group of researchers from the University of California at Davis analyzed 399 blood serum specimens taken from women who gave birth in the Bay Area in the 1960s. The study group was comprised of women from families who were enrolled in a Kaiser Permanente health-care plan. They all came from families with stable employment, thus excluding low-income mothers, who might be more likely to have problems with nutrition and alcohol and drug use. What the investigators found in their scrupulous study was that the women who had the greatest body burden of PCBs were 33 percent less likely to give birth to male children than the least exposed women. In a grim testament to the power of even the minutest amounts of the chemical to skew sex ratio, the women most exposed to PCBs were 7 percent less likely to bear a male child for every *one* part per *billion* of PCBs in their blood.

Dr. Irva Hertz-Picciotto, the lead author of the seminal study, opined that she wasn't sure exactly how PCBs were doing their damage, but it appeared that they either compromised the ability of sperm to fertilize the egg or, more likely, somehow caused a higher number of spontaneous abortions of male embryos than of females. Whatever the mechanism, Hertz-Picciotto was convinced by the voluminous data collected and analyzed that PCBs or their metabolites were the culprits.

Perhaps the most speculative, frightening, and complex of all research linking PCBs to catastrophic childhood illness involves autism in industrialized nations. In the United States there has been a staggering tenfold increase in the number of identified cases of autism in less than twenty-five years. A recent study found that one in one hundred American children suffers from an autism spectrum disorder. But even though disease rates are soaring statistically, there are some researchers—with valid arguments—who say that there is no "autism epidemic," only an increase in diagnosis. But such a stance misses the point. Whether or not the rates are actually increasing—by any measure—autistic disorders are the most prevalent and devastating developmental/behavioral disease of childhood (and later adulthood) in the Western world. And it is this fact that has scientific investigators looking for environmental triggers. As one neuro-researcher put it, "the only way you can explain [it] is if there are environmental factors that are strongly expressed and relatively widely distributed in the environment."

Taking a contrary position, chemical industry apologists in the scientific community preach that the etiology of autism must be almost completely genetic because identical twin studies have shown that more than 95 percent of the time, when one twin becomes autistic, the other will as well. However, on simple examination the theory of overwhelmingly genetic causation of autism is easily debunked by two factors. First, both twins could have been exposed to the same chemical agents in utero. And secondly, autistic adults—especially when they are highly autistic—are less likely to have families, making them, as a genetic group, less likely to pass along their "autistic" genotype. In other words, autism is not a successful human trait, since it is suppressed by the most powerful of genetic forces—that of Darwinian selection.

A more likely scenario is that the genetic predisposition for autism is fairly widespread in the human population—but it takes a *trigger* to make the illness full blown. In fact, scientists have been aware for decades that exposure to certain synthetic chemical substances during pregnancy can cause autism. Approximately one out of three children whose mothers were prescribed thalidomide (a drug given to pregnant women in the 1960s to alleviate morning sickness) became autistic if the women took the drug between the nineteenth and twenty-fourth days of pregnancy. Researchers also know that it is during this precise window that the brain of the fetus undergoes a major transformation.

Even more compelling, recall that PCBs are believed to strongly and adversely affect the immune systems of children exposed in utero or perhaps via breast-feeding. This provides a disturbing link to the mysterious and calamitous tale of autism. Autistic children have comparatively high rates of immune disorders—they are prone to infections, food allergies, and gastrointestinal problems. In fact, many autistic children have abnormal immune systems, according to studies done at the UC–Davis MIND Institute. Scientists there found what they characterized as clear differences in cellular immune responses to bacterial agents between autistic children and children who were developing normally.

Even though there isn't anything even close to a smoking gun yet discovered that would point to a relationship between PCBs and autism, researchers have performed tests on lab rats that portend a definitive link. Neuroscientists at the MIND Institute have replicated auditory problems in test animals exposed to extremely small amounts of polychlorinated biphenyl that resemble hearing dysfunction found in many children with autism. Dr. Isaac Pessah, a lead autism researcher with the institute, theorized that PCBs were causing the brains of rats exposed to the chemical during critical windows of development to become miswired for sound processing.

But questions about the relationship between PCBs and autism will probably not be decisively answered anytime soon. According to Dr. Pessah, since PCBs come in so many congeners, and since all of them can have their toxicity altered by age, exposure to the environment, and how they have been metabolized, even the most advanced analytical techniques may not be capable of accurately measuring the true exposure rates of autistic children to PCBs, especially if they occurred prenatally.

If determining the link between PCBs and autism is a scientifically tantalizing problem and an extremely urgent one to solve, then pinpointing the triggering mechanism that may be involved is an even more pressing puzzle. In this regard, one of the most promising lines of research appears to be the powerful negative effect that PCBs have on a protective enzyme called glutathione. Laboratory experiments have shown that PCBs can cause an increase of "oxidative stress" in cellular activities associated with the brains of mammals. Oxidative stress, in simple terms, is a destructive cellular activity that produces harmful free radicals—highly reactive chemical agents that are believed to cause molecular damage in susceptible brain tissue.

Oxidative stress can be created by a variety of toxic insults, including air pollution, smoking, pesticides, food additives, and industrial chemicals. The type of damage caused by oxidative stress via free radical production varies, but perhaps the most dangerous changes that free radicals induce are those to DNA, the cellular material that contains all genetic information—the blueprints of life. In the case of autism, it is believed that free radicals may damage how DNA controls the replication of neurons, or brain cells.

To offset free radical damage, human beings and other mammals deploy antioxidants—substances, generally enzymes, that the body manufactures to protect cells from damage. Glutathione is one of the most powerful antioxidant enzymes that guard human beings against the production of free radicals and the ravages of oxidative stress—and PCBs are widely known by scientists to be potent antagonists (destroyers) of glutathione.

So what are the clues in this detective story? Where is the connection between PCBs, glutathione, and autism? Actually there are more than one. First is the fact that boys account for 70 percent of all autism cases, and research indicates that boys are born with generally weaker antioxidant capacities than are girls. (Estrogen can, at times, be a potent antioxidant and is known to give substantial protection against free radical damage to young mammalian females.) As we've just seen, studies have revealed that a common trait in persons with autism, besides the obvious behavioral elements and physical auditory problems, is their impaired or altered immune systems, and PCBs are known to cause serious damage to the immune systems of children—a fact known since the Yusho poisonings in the late 1960s.

Isaac Pessah is also studying another venue that allows neurotoxic damage by PCBs to occur—also with potential links to autism. He and his colleagues exposed rats to extremely low doses—comparable in weight and amounts to what human infants might be exposed to—of two types of PCBs (PCB95 and PCB170) by feeding them PCB-laced cookies. By using a series of water maze tests, Pessah found that the rats dosed with PCBs had brain circuitry in the hippocampal region that was abnormally excited compared to that of control animals. As to a link with autism, Pessah explained, "We think that in autism, for example, at-risk children have deficient inhibitory circuits. So, if you have a PCB that promotes the excitatory side of the circuit, they would be much more at risk of developing the disorder."

Another study of PCBs and neurotoxicity, coauthored by Pamela Lein, also of UC–Davis, and Dongren Yang, found that PCBs significantly deformed the growth patterns of dendrites (brain nerve cells) and their subsequent plasticity—the malleability of the nerve cell—in the rats exposed to PCBs. These animals also took markedly longer to master water maze training than their cohorts who were not exposed to PCBs.

"This tells us that PCBs are altering dendritic growth and plasticity," Lein said of her experiments. "The results are important because problems in dendritic growth and plasticity have previously been implicated in many neurodevelopmental disorders, including autism, schizophrenia, and mental retardation. Dendritic plasticity is important to how we process information and, when you perturb that, you interfere with complex behaviors like learning and memory."

Does this mean that PCBs are the main causative agent for autism? In some cases they could be, and in others, probably not. Regardless, as research goes forward, it's a good bet that industrial chemical contamination will be found to play some role—perhaps a pivotal one—in the etiology of autism. However, along with the environment of those affected by this often terrible and complex disease, genes must also play an important role. Understanding this is simple: environment—exposure to pollutants, diet, and such—cannot be totally to blame in causing autism. Almost all children in industrialized nations are exposed to some ambient levels of known neurotoxicants like PCBs and mercury, but most do not become autistic. So how each of us is genetically tuned to deploy glutathione and other antioxidants certainly would appear to be important . . . but as important as environmental factors? Nobody knows.

What is known about the impacts on children of PCBs and their toxic cohorts—at least according to a scientific review by two of the world's leading neuroscientists, Dr. Philippe Grandjean of the Harvard School of Public Health and Philip Landrigan of the Mount Sinai School of

Medicine in New York—is that industrial chemicals are causing "a silent pandemic" of brain damage to millions of children worldwide.

In an article published in the British journal *Lancet* in the fall of 2006, Grandjean and Landrigan studied five neurotoxicants and their effects on brain development in children. Among the five, the authors included PCBs as a strong candidate for contributing to the meteoric rise of autism, probably through some synergistic reaction with over two hundred of the known industrial chemicals capable of causing permanent brain damage in children.

The landmark study by Grandjean and Landrigan was inexplicably released by the Harvard School of Public Health on Election Day 2006, when it virtually died in the crammed news cycle. Nonetheless, Dr. Grandjean's words in the press release, although almost completely ignored by the media, were an immensely important and unusually eloquent plea: "The brains of our children are our most precious resource, and we haven't recognized how vulnerable they are. We must make protection of the young brain a paramount goal of public health protection. You have only one chance to develop a brain."

CHAPTER 7

Adult Realities

The chemical threat is the ultimate threat to mankind—worse than bombs and war. You cannot hide from it. It reaches everywhere in the world.

INGMAR, INUIT TRIBAL ELDER,

FROM *Silent Snow*, BY MARLA CONE

There are hundreds, perhaps thousands, of studies involving PCBs and their effects on adult human beings. However, if you wanted to find a paper that proved beyond any doubt that background levels of PCBs were the direct cause of illness in adults—among all the research over the last forty years—you wouldn't find a single one.

Still investigators have reasonable ways to point an accusatory finger at PCBs. There is a scientific philosophy that is used to make judgments about the toxicity of substances—particularly scientifically complex industrial chemical pollutants like PCBs—and it is called the "weight of evidence" approach. The meaning is straightforward: if there are enough studies that link a substance such as PCBs to negative health effects or ecological damage, then the substance can be thought of scientifically as a toxin—regardless of a lack of proof in any single study.

So, if we take into consideration the current analytical shortcomings of the biological sciences, but we apply the weight-of-evidence principle to PCBs, we find that—as was suspected back in the 1930s—PCBs probably cause cancer in genetically susceptible adults through environmental exposure. But, just as there never was one for tobacco and cancer causation, there is no smoking gun that can prove to everyone that PCBs cause cancer in human adults.

At least one reason for this lack of smoking-gun evidence involves a grand irony. Human beings around the globe are so contaminated with industrial chemicals and their metabolites that it is virtually impossible to pinpoint which chemical is carcinogenic on its own. Then add to this problem the most hotly contentious biological brouhaha of the last hundred years—the "genes versus environment" dispute.

Researchers who see rising cancer rates argue logically that the increases—sometimes in epidemic numbers, as with testicular cancer —must be somehow tied to changes in the environment, since our pre-industrial forebears didn't develop cancer at the rates that humans are now experiencing. Conversely, chemical industry apologists argue that the cancer rates are rising, for the most part, because of better detection and longer life spans.

In actuality, the dispute has been scientifically settled, even if chemical-industry-oriented scientists are generally reluctant to accept it. According to the most thorough research ever undertaken on the subject, a person's environment—lifestyle, occupation, number of infections experienced, and diet—causes a high percentage of cancers. Environmental exposures, on average, will cause two, three, or four times the number of cancers as do our genes.

The basis for this finding, which has been called the "gold standard" for cancer causation, is the Scandinavian twin studies. Researchers in the late 1990s looked at a total population of almost ninety thousand twins in Scandinavia, including ten thousand cases of cancer in the study group. What they found was that, of the major cancers, between 60 and 80 percent were caused by environment—not heredity. Another perspective on this finding is that though the victims may have been genetically susceptible to a specific type of cancer, they wouldn't have developed cancer if it were not for their exposure to a catalytic agent in their environment.

For women, cancers of the reproductive organs—cervix, uterus, and ovaries—were all approximately four times more likely to be caused by environment than genes. In the case of uterine cancer, environment was eight times more likely to be the cause. In cancer of the cervix, the researchers found a pure heredity factor of zero. Otherwise stated—*all* cervical cancers were caused by the environment, probably through viral or bacteriological triggering. For men, environmental factors were the cause of about 60 to 65 percent of the different cancers in the twins who were studied.

Other comprehensive research supported the link between the environment and cancer. The most convincing was a study done of rural Asian women who had recently migrated to the United States. Their breast cancer rates were 80 percent lower than the breast cancer rates of third-generation Asian American women with similar genetic backgrounds. Since lifestyle factors such as alcohol consumption and tobacco use are not generally thought to be major causes of breast cancer, the difference for these women was their environment and, most particularly, it was thought, some aspect of their westernized diet.

So how do PCBs factor into the environmental etiology of cancer? Even though polychlorinated biphenyl is almost always the first or second most common toxic industrial chemical found in any given human being, it probably is selective in the types of cancer it might trigger. Besides breast cancer in women with a vulnerable genetic makeup (as we will find in the next chapter), the adult cancer linked most strongly to PCBs is non-Hodgkin's lymphoma, a family of cancers that involve lymph nodes, a component of our immune system.

The rate of increase for non-Hodgkin's lymphoma in the United States has been frighteningly steep since the end of World War II. In Caucasian men, the annual incidence of the disease increased by almost 250 percent from 1950 to 1991. This made the growth rate for non-Hodgkin's lymphoma the third fastest for all male cancers (behind prostate cancer and malignant melanoma) during the last half of the twentieth century. The rate has leveled off in industrialized countries as of the beginning of the millennium, but it is not decreasing.

While AIDS, organ transplants, and some uncommon genetic disorders greatly increase the risk of developing non-Hodgkin's lymphoma, those risk factors do not come close to explaining the increased occurrence of the disease—not only in America, but around the globe. Researchers faced with finding out what caused this once relatively rare cancer to become so prevalent thought that it might be tied to the use of DDT, since the banned pesticide was known to interfere with the immune systems of a variety of laboratory test animals.

But in the mid-1990s, after a thorough investigation into the connection between non-Hodgkin's lymphoma and DDT, Dr. Nathaniel Rothman of the National Cancer Institute found that there was no link with the pesticide or other pesticides that he tested. However, there was a strong association between non-Hodgkin's lymphoma and PCBs in the records of the twenty-six thousand participants in the study.

Of even more concern, the levels of PCBs in the blood of the non-Hodgkin's lymphoma victims were not that much greater than the average body burden of PCBs in adult Americans, even though they were more than four times more likely to get the disease. And the fourfold risk factor was for victims who *hadn't* been exposed to PCBs at work.

Dr. Rothman's extensive study also had another startling finding. Rothman and his colleagues discovered that if victims of non-Hodgkin's lymphoma had been exposed to the Epstein-Barr virus (a common herpes-type virus) and had moderately higher levels of PCBs in their blood, the risk of developing non-Hodgkin's lymphoma was magnified more than twenty times. Somehow, through some sort of synergy, the Epstein-Barr virus greatly amplified the carcinogenicity of PCBs, or possibly the reverse.

Though the relationships among Epstein-Barr, non-Hodgkin's lymphoma, and PCBs were surprising research outcomes, experts weren't necessarily shocked. The results were in line with long-held knowledge that PCBs have a strong suppressive effect on the immune system of almost all species of animals. And PCBs were known by investigators to act synergistically with other industrial chemicals as well, making any mix theoretically far more toxic. However, Dr. Rothman's findings were the first to document a synergy between industrial chemicals and a virus—a very disconcerting discovery to be sure.

In another comparatively strong body of evidence that PCBs can

cause cancer in adult humans, epidemiologists found a correlation between occupational exposure to PCBs and malignant melanoma. In one
of the largest PCB-related investigations ever undertaken, researchers at the University of North Carolina at Chapel Hill studied nearly
140,000 workers who were longtime employees of power companies.
Their findings, published in 1997, "suggested" that PCBs did indeed
induce cancer, and malignant melanoma was of greatest concern. (An
earlier study conducted in 1992, although not as comprehensive, re-
flected much the same result.)

The North Carolina study also indicated that occupational exposure
to PCBs increased the chance of brain cancer among workers. In 2006,
the National Institute for Occupational Safety and Health conducted
research into the relationship between brain cancer and PCBs at a large
capacitor plant in Indiana where PCBs had been used for decades in
manufacturing processes. Here too an association between brain cancer
cases and PCBs was found in the workers. But the authors of the study
issued a caveat: there was no demonstrable exposure–response relationship. In other words, the levels of PCBs found in the workers' blood
didn't correspond to the incidence of cancer.

Whatever the outcome of the debate surrounding PCBs and cancer may
be and whatever headlines it may someday generate, the most consistent
and disastrous impact that PCBs are believed to have on adult humans,
other than cancer, is on the brain in the form of Parkinson's disease.
Parkinson's disease generally strikes older people and is a progressive
degenerative disorder that causes tremors and destruction of neurons. In
the United States there are approximately 1.5 million persons with the
incurable disorder.

While a person's genotype is thought to play some part in the etiology
of Parkinson's disease, according to an American study of nearly twenty
thousand twins, the primary cause of the disease is exposure to some
agent in the victim's environment or diet, and PCBs are a major suspect.
Two studies by scientists at the University of Rochester, published in
December 2004 and February 2005, demonstrated that when brain cells
are exposed to PCBs (and some pesticides), they are left vulnerable to

molecular injury that may produce catastrophic results. According to the research team led by Dr. Lisa Opanashuk, PCBs—even in low parts per billion—do their primary damage by disrupting the functioning of neurons sensitive to dopamine. (Dopamine is a neurotransmitter—a chemical messenger—that plays a key role in how the brain controls body movement and balance, as well as in emotional responses, including the ability to experience normal amounts of pleasure and pain.)

More subtle neurological damage than that of Parkinson's can occur from exposure to PCBs. Two well-executed Canadian studies in the late 1990s of men and women who consumed fish known to be contaminated with PCBs from the St. Lawrence lakes region showed that they had significantly slower motor responses, higher rates of mental confusion, and poorer results on certain memory and attention tests. What made the study so credible was that the researchers vigilantly factored in adjustments for age, education, and alcohol intake—leaving PCBs as the most prominent offender.

U.S. scientists conducted similar studies on sport fishermen who regularly ate fish caught in the Great Lakes—generally second only to Baltic-region fish in heavy PCB contamination. The investigators focused on anglers who consumed twenty-four pounds or more of Great Lakes fish per year. Most were found to have relatively high levels of PCBs in their blood. Impairments of short-term memory were clearly evident in the fish eaters. It was also reported that their word-learning test results were well below normal.

As these words are written, if you Google the words "PCBs" and "immunity" you get 503,000 hits. Evidently, research into the question of the relationship between immunity and toxic substances—industrial chemical contaminants like PCBs—is a contemporary scientific priority. And well it should be, because immune response is a life-or-death issue for most animal species, including us.

Regardless of the difficulties in gauging their precise influence on adult human immunity, PCBs are of special importance for two reasons: first, because researchers have demonstrated in numerous lab experiments that they have a strong capacity for interfering with

the delicate chemical interactions that coordinate the proper functioning of all vertebrate immune systems; second, because PCBs are often
the most ubiquitous immunotoxic agent in any given part of our
biosphere.

Swedish scientists addressed the link between PCBs and immunity
in another of their vanguard PCB studies—this time of fishermen who
regularly ate fish with high fat content caught in the Baltic Sea. After analyzing various parameters of the men's immune systems—white
cell count, lymphocytes, and immunoglobulin levels—the researchers
found a significant correlation between suppressed immune reactions in
their subjects and the amount of fish, especially salmon, they consumed.

However, the study was weakened by the fact that the men's exact
body burdens of PCBs, either in fatty tissue or blood serum, were not
known. Consequently the suppression of their immune systems might
have been caused by other contaminants in the fish, or, for that matter,
by PCBs synergistically reacting with other industrial chemicals.

This was not the case in studies of adult Japanese victims of Yusho
poisoning and their immune response to high PCB exposure through
ingestion of contaminated cooking oil. Symptoms of immunosuppression were consistently observed over a multiyear period, with the Yusho
victims suffering from an increased incidence of respiratory illnesses
and infections that were linked to damaged immune systems.

Perhaps the most advanced research into the effects of PCBs involves
their binding to critical genes. Dr. Bruce Blumberg and his team at University of California at Irvine have found that PCBs can significantly
impact the ability of our body to detoxify other industrial chemicals
and naturally occurring toxins to which we are exposed every day of
our lives.

According to his discovery of, and research into, a gene called the
SXR receptor—a critical evolutionary element in our immunological
makeup—PCBs have the potential to drastically alter our bodies' natural protective responses to environmental contaminants. Dr. Blumberg
found that a nearly omnipresent and particularly toxic variety of PCBs
with higher chlorine content demonstrated a strong potential for harming the human immune response through SXR mechanisms. But as

usual with the mysterious "magic fluid," not all 209 PCB varieties exhibited the destructive behavior.

A somewhat less specific but no less damning assessment of the relationship between PCBs and autoimmune disease in adults comes from Texas A&M University insecticide toxicologist Dr. Frederick W. Plapp. He sees a definite connection between the large increase in autoimmune diseases in recent decades and human exposure to PCBs, as well as other industrial chemicals. Plapp believes that a list of environmentally induced autoimmune illnesses should include fibromyalgia, chronic fatigue syndrome, multiple chemical sensitivity, and Gulf War syndrome. Plapp also judges that other autoimmune diseases such as rheumatoid arthritis, insulin-dependent diabetes, systemic lupus, and multiple sclerosis may be triggered by toxins in the environment of the genetically predisposed victim.

Giving a possible explanation for a mechanism that would link all the diseases, Plapp concludes that PCBs, and perhaps other industrial chemicals, "poison" a carrier protein called transferrin, a substance produced in the endocrine system of most mammals. Transferrin helps provide vitamin A to organs that need the vitamin to function properly. Researchers have long speculated that vitamin A deficiency is a precursor to the development of many, if not all, autoimmune diseases in vertebrates—as we'll find out in chapter 9.

But perhaps the most stunning—and reasonable—speculation by investigators surrounds the possibility that industrial contaminants such as PCBs and their cohorts, dioxins, are contributing to, if not causing outright, the global epidemic of obesity. Researchers led by Dr. Jerome Ruzzin, a Norwegian scientist, published a paper in early 2010 establishing for the first time that PCBs and other persistent organic pollutants, known as POPs, cause insulin resistance in mammals. This coupled with the fact that insulin resistance can lead to type 2 diabetes, now blossoming in human populations in epidemic proportions, made the Ruzzin study a potential smoking gun linking PCBs—the most widely distributed and persistent POP in the biosphere—and obesity in genetically susceptible human populations.

As more studies are undertaken regarding the link between obesity and POPs, it's unlikely that scientists will be able to prove to the satisfaction of the chemical industry apologists that there's a direct link. Experiments using human subjects are out of the question, and even if they were done under some Nazi-type research regimen, they still would probably prove nothing beyond a doubt. There's just too much complexity, and too many industrial chemicals that contaminate human beings, to tease out one or two as the causative factors for type 2 diabetes. It looks like we simply must accept reasonable proof—that of the weight of evidence.

If the "environmental trigger" concept is correct—and there is no reason to believe that it isn't—it means that if someone is not exposed to a triggering agent such as PCBs, a specific disease will not develop. In other words, without a trigger being pulled, no bullet is fired. Even though a person might have a potentially lethal genetic weakness that would make them prone to develop an autoimmune disease, or perhaps cancer, that flaw might always remain latent without exposure to a catalyst. PCBs (and other industrial chemical pollutants) very likely *allow* some specific diseases—such as respiratory infections—to occur even though the chemicals themselves aren't the actual *cause* of the disease.

This triggering behavior by PCBs and their toxic cohorts skews the traditional way of viewing toxicity, since there isn't necessarily a direct cause-and-effect relationship between the chemical and the disease. While this might sound like a rather minor academic exercise, in fact the practical impact is huge. It means that in order to really protect ourselves from diseases that are *triggered* by PCBs and their chemical cousins—not necessarily *caused* by them—we have to scientifically, legally, and politically rethink just what toxicity is.

CHAPTER 8

PCBs, Breast Cancer, and Hidden Agendas

Chemophobia, the unreasonable fear of chemicals, is a common public reaction to scientific or media reports suggesting that exposure to various environmental contaminants may pose a threat to health.

DR. STEPHEN SAFE, *The New England Journal of Medicine*

PCBs have been found everywhere on the planet, in the deepest ocean trenches and the highest mountain ranges. So efficient are PCBs at migrating from the environment to the cells of living creatures that there is probably not a human being alive who doesn't have PCBs locked somewhere in his or her tissues. We have all been chemically tattooed. We are all participants in the largest involuntary lab test in human history. Consequently, every person reading these words is directly or indirectly part of a heated scientific controversy: are PCBs contributing to the epidemic of breast cancer in the United States and Europe?

In fact, the question had been answered ten years earlier. Perhaps the foremost PCB researcher in the world declared that PCBs were not harming women. He said essentially that PCBs were safe, but their reputation had suffered from an unfortunately sensationalistic media. The announcement of the safety of PCBs came from a scientist at Texas A&M, implausibly named Stephen H. Safe. Dr. Safe was (and arguably still is) the world's preeminent expert on the subject of the toxicity of PCBs. For thirty years, he had investigated their chemistry as thoroughly as anyone on the planet and his work was cited thousands of times, more than any other scientist, living or dead. If anybody could pronounce PCBs safe to human health, it was Dr. Stephen H. Safe.

In his exoneration of PCBs, Dr. Safe pointed to the largest-scale research project of its time. The results, published by the ultra-prestigious *New England Journal of Medicine* in the fall of 1997, had failed to find any connection between PCBs and breast cancer in hundreds of women who were chosen as test subjects. In fact, according to some interpretations, the research indicated that women with higher body burdens of PCBs actually had *lower* rates of breast cancer.

Dr. Safe was given a full editorial page by the *New England Journal of Medicine* to expand on his views of the unnecessary fears regarding PCBs specifically, as well as other trace industrial chemicals that are found in everyone. Coining a new term, Safe called the public reaction to scientific and media reports about PCBs "chemophobia," a sort of modern-day hysteria that mostly affected women—or such was his implication.

The editorial got strong positive coverage in both the *New York Times* and the *Wall Street Journal*. Both hailed the breast cancer study and Dr. Safe's views as illuminating and authoritative. The *New York Times*'s science reporter summarized Safe's editorial and the breast cancer study by writing, "One more environmental scare bit the dust last week as scientists from the Harvard School of Public Health reported that their large and meticulous study found no evidence that exposure to the chemicals DDT and PCB's [*sic*] are linked to breast cancer."

Dr. Safe was quoted as telling the *Times* reporter that it was time to stop trying to make a connection between breast cancer and synthetic organochlorines like PCBs. "For advocates [of the idea] it's never ending. But for other people, there may be times when we want to spend our money on other things," said Safe. He opined that the public just had to move on.

But it wasn't time to move on. Dr. Safe would be shown to be wrong—perhaps dead wrong—both in his quasi-political pronouncements and his scientific analysis of the safety of PCBs. His controversial scientific judgments would encompass the most profound health concerns of more than twenty-three million females in the United States and hundreds of millions of women worldwide, making a tragically fascinating and ultimately disheartening tale about the realities of gender politics and the influence of money on science.

In contrast to the link between PCBs and testicular cancer, there is no smoking-gun relationship between the PCBs and breast cancer. However—since the turn of the millennium—a persuasive new body of evidence has been compiled, mostly by female scientists, that argues that PCBs present a quite serious breast cancer risk to women in America and worldwide.

The study that Dr. Safe used to absolve PCBs of causing breast cancer was done by lead researcher Dr. David J. Hunter, of the Harvard School of Public Health. Dr. Hunter, an Australian, was the Vincent L. Gregory Professor in Cancer Prevention, an endowed chair that gave him a prestigious academic platform he could use to advance his views on the causes of breast cancer.

Dr. Hunter's study was published in the fall of 1997 in the *New England Journal of Medicine*, along with the "chemophobia" editorial by Dr. Safe, which declared that PCBs were irrelevant to the breast cancer epidemic. Hunter's study of PCBs and DDT was comprehensive—at least as far as numbers were concerned. Using data from the famous Nurses' Health Study, which had stored thirty-two thousand blood samples from nurses across the country, Dr. Hunter and his colleagues chose 240 women who had developed breast cancer and then found a matching number of nurses who had not. After analyzing the research, the Hunter team found that there was no relationship between breast cancer and the amount of PCBs in the blood of the nurses. The levels of PCBs in both groups were essentially the same.

It all seemed so clear-cut. Another chemophobic myth had been debunked. Or had it? As lead researcher, Dr. Hunter—with Dr. Safe's editorial endorsement—had apparently based his breast cancer study on the premise that all women are genetically identical when it comes to how their bodies deal with industrial chemicals like PCBs.

There were, of course, obvious benefits to the simplicity of Dr. Hunter's thinking. It certainly made for an easy study. If all women were basically the same genetically for the purposes of the research, then all you had to do was locate some with breast cancer and compare their blood serum levels of industrial contaminants such as PCBs with the levels in women without breast cancer and voilà! If the levels were about the same in both groups of women, PCBs couldn't be the cause of breast cancer in women.

In retrospect, lumping all women together as Dr. Hunter did, believing that they all would have the same genetic response to chemical contaminants—ignoring the possibility that racial or ethnic subgroups of women such as African Americans or Ashkenazi Jewish females, for example, might have dissimilar genetically based reactions to toxins (which they indeed do)—would seem implausible behavior for an epidemiologist with Hunter's reputation, and implausible for Safe to accept. But they apparently did just that.

Four female scientists were named as coauthors of Dr. Hunter's "Not-to-Worry" study along with five males. Surprisingly, only one of the women publicly questioned Hunter's conclusions, and her dissent was meek and muffled. She told inquiring journalists that it might be "premature" to clear PCBs and other organochlorines of increasing the risk of breast cancer, because of the possibility it "might be important for some groups of women."

Other women scientists—mostly older and familiar with the controversy over PCBs and breast cancer, but not involved directly in the study—were deferential toward Dr. Hunter and his findings. One even told a reporter that Hunter's research had such a "strong design," that it pretty much negated any real possibility PCBs were associated with breast cancer.

None of the male coauthors of the Hunter Study—most of whom were middle-aged and perhaps a lot more concerned about their arteries than the chance of developing breast cancer—publicly questioned the conclusions and opinions of their eminent colleagues. It would take a younger scientist—one with something to lose—to do that.

• • •

The first researcher to look at the real-world nuances of the relationship between PCBs and breast cancer was a German-born scientist with the State University of New York at Buffalo, Dr. Kirsten Moysich. Having received her PhD only two years earlier, Dr. Moysich was relatively young for the intricate epidemiological field that she had chosen for her first major research effort—breast cancer etiology and organochlorines—but she was exceptionally bright and energetic, and she had a hunch.

The hunch was in fact pretty logical. She thought that it was unlikely that a woman's genetic makeup—her genotype—would not have any bearing on increased risk of breast cancer if she had higher exposure levels of PCBs. While certainly a reasonable gut feeling for Dr. Moysich, it nevertheless could be viewed as a long shot for a scientist in the process of launching a research career. Not only was she challenging the understanding of two very powerful members of the scientific establishment, Drs. Hunter and Safe, but other studies had found no relationship between breast cancer and blood serum levels of PCBs.

But Dr. Moysich was not about to deny her suspicion. Moysich located more than one hundred postmenopausal women with breast cancer from a previous study group in western New York State. She also got a like number of controls, postmenopausal women without breast cancer, to participate in her study. All were given blood tests and then classified by genotype. It was a first for research on PCBs and breast cancer—and the results were nothing less than stunning.

While the overall risk of breast cancer was not linked to PCB levels, the women in the study who carried a deleterious version of a gene that helps determine how the body metabolizes toxic substances—called the CYP1A1 gene—had almost three times the risk of developing breast cancer if they carried somewhat elevated PCB levels in their blood. Contrary to the conclusions drawn so dramatically and confidently by Drs. Safe and Hunter, Moysich had discovered a solid link between PCBs and breast cancer, but *only* for women who had a specific genotype.

By 2002, three years after the groundbreaking Moysich study, another female researcher, with a surprising job position, replicated the findings of Dr. Moysich nearly precisely—an almost unheard-of event in large-scale research studies. Dr. Francine Laden was an epidemiologist with the Harvard School of Public Health—the very same institution where Dr. Hunter worked. Dr. Laden never publicly dissented from the editorial conclusions that Dr. Safe drew about the lack of any relationship between PCBs and breast cancer. However, she used the very same study group—the Nurses' Health Study—for her research. What Laden and her team found was that postmenopausal women carrying only somewhat elevated blood levels of PCBs had—just as in the Moysich findings—almost exactly a threefold risk of getting breast cancer if they carried the bad type of the CYP1A1 gene.

Dr. Laden's work was good, but not definitive. There were areas she and her associates didn't explore that were pertinent to the PCB–breast cancer connection. One of the problems in studying PCBs in humans, or any living creature, is that PCBs come in so many different flavors—209 varieties with very similar molecular structures—each of which, although almost chemically identical to the others, can do *very* dissimilar things within the human body. Dr. Laden had speculated that some of the varieties of PCBs might be more likely to affect breast cancer risk in women who had the potentially dangerous CYP1A1 gene, but she never fully investigated the possibility.

It would take yet another female scientist to make the link between PCBs and breast cancer in genetically susceptible women obvious for all but the pathologically skeptical or—as we will see—those with a financial or personal interest in disbelief.

The most thorough examination and subsequent demolition of the "PCBs Don't Cause Breast Cancer" editorial was published in June of 2004 in the *American Journal of Epidemiology*. It certainly was quite an honor for Dr. Yawei Zhang, a surprisingly youngish woman researcher from mainland China, where she had received her medical degree before finding a scientific home at the Yale School of Public health. Within two years of earning her PhD from Yale, Dr. Zhang had been author or co-author of forty-three peer-reviewed research papers. Most involved the elaborate dynamics of genes and cancer causation, and had quickly established her as one of the premier epidemiologists in the world on the subject.

Dr. Zhang's *American Journal of Epidemiology* study was beautifully executed. She had methodically tied up all the loose ends of her predecessors' studies. She not only analyzed the genetic subgroupings of the women in her study—some eight hundred women with and without breast cancer—but she also scrutinized the most complicated area of the link between PCBs and cancer, the individual roles that nine different varieties of PCBs played.

After analyzing how the women reacted to the specific varieties of PCBs, Dr. Zhang double-checked the accuracy of her data on the blood serum levels of PCBs in her subjects by comparing them to fatty tissue samples that they had donated as well. Cross-checking all her data was not only costly in terms of funding, but was complicated and time consuming. But Zhang was aggressively scrupulous and her efforts were to yield undeniable results.

Dr. Zhang confirmed that the earlier studies were correct. She also ascertained that their findings were probably on the low side for breast cancer risk from PCBs. Zhang and her colleagues found that women who carried the bad CYP1A1 gene had almost *four* times the risk of developing breast cancer if they had higher body burdens of PCBs than women who didn't carry the gene. And with certain varieties of PCBs, known to be more toxic, women had almost *five* times the likelihood of getting breast cancer if their body burdens of these types of PCBs were on the upper end of the scale and they had the bad CYP1A1 variant gene as well.

But what was the significance of all these findings? How was all this exacting science applicable to the real world, and pollution, and people getting sick and dying from cancer? Actually the investigations of researchers Moysich, Laden, and Zhang—like Hardell's work with PCBs and testicular cancer—will probably change the way all cancer research is done.

It turns out that approximately 15 percent of all females in North America carry the bad CYP1A1 variant gene. With a population of about 300 million people in the United States as of 2010 this would mean that roughly twenty-three million females, young and old, are at higher risk, sometimes much higher risk, of developing breast cancer if they have even mildly elevated body burdens of PCBs and perhaps other industrial chemical contaminants.

So was there a moral to the story? Certainly. Maybe lots of them. But one for sure: Dr. Safe's legions of hysterical "chemophobic" women probably had good reason to be hysterical, and chemophobic too, not only because of the health threat, but because the science of the PCB apologists was thoroughly tainted with hidden agendas—financial, personal, and perhaps gender-based as well.

Dr. Safe followed up his "PCBs Don't Cause Breast Cancer" editorial in the *New England Journal of Medicine* with a series of interviews, telling reporters that those who questioned the safety of PCBs and other chemical pollutants were practitioners of "paparazzi science." It was big news for a media hungry for any sort of public trashing between professionals, especially one emanating from the usually sedate halls of the scientific establishment. But in an apparent reversal, Safe seemed to think that the alarm of the "paparazzi" scientists wasn't exactly worthless, "but, for the most part, I think the concern is overblown. I wouldn't dismiss it, but I'm hopeful that the concerns are not a major health concern. At least in term of humans."

Yet within a year or so, Dr. Safe was back to what some would call a mass denigration of multiple studies that linked PCBs and other synthetic organochlorines with a variety of human ills. Never acknowledging the thousands of Asian victims of Yusho and Yu-Cheng, Safe said that he didn't believe there were any "identified linkages" between human disease and contaminants such as PCBs.

Naturally, Dr. Safe's well-publicized views were an uncommonly good public relations windfall for the chemical industry, under attack by conservationists since the publication of Rachel Carson's *Silent Spring*. The American Chemistry Council, a PR arm of chemical manufacturers, seized on Safe's exoneration of PCBs as causing breast cancer—or any human health problems—with a lengthy "report" to the public titled "Chemicals in the Environment and the Endocrine System." Citing him fourteen times, the publicists made Safe the veritable "Dr. Feelgood" for industrial pollution.

• • •

For chemical manufacturers to use Dr. Safe's work to counter bad media coverage certainly made sense since—according to published and broadcast reports—they were directly or indirectly funding the Texas A&M professor with hundreds of thousands of dollars, perhaps millions in total over the years (in current dollar terms). To the intense embarrassment of the *New England Journal of Medicine*, an environmental group tipped off the *Boston Globe*'s environmental investigative journalist, Larry Tye, to Dr. Safe's financial arrangements with the chemical industry.

According to Tye's reporting—later confirmed by Dr. Safe himself in an interview with Doug Hamilton of PBS's *Frontline*—in the five years prior to his editorial, the chemical industry had financed approximately 20 percent a year of Safe's research at Texas A&M. In the three years preceding his headline-making editorial absolution of PCBs, Dr. Safe's lab work was being financed by the Chemical Manufacturers Association to the tune of $150,000 annually, along with funds from the Chlorine Chemistry Council and the Cattlemen's Association.

His editors at the *New England Journal of Medicine* must have been mortified at Tye's revelations in early 1998. The *Journal* had some of the toughest—if not *the* toughest—conflict-of-interest rules for authors of any medical publication in the United States. Eight years earlier, rather than just requiring authors to disclose potential conflicts of interest as most other reputable medical periodicals did, the *Journal*'s guidelines began obligating editors to *reject* all articles or editorials from any author connected to entities with financial interests in the issue being discussed.

The *Journal*'s editor-in-chief, Dr. Jerome P. Kassirer, admitted that he didn't know about Dr. Safe's connections with the chemical industry, but he said he probably would have published the editorial anyway because Safe's funding by chemical interests had ended five months earlier. But for some, Kassirer's excuse didn't seem well founded, because Safe had almost certainly written the editorial—or formed his opinions, at the very least—while he was receiving six-figure funding from the Chemical Manufacturers Association. (Be that as it may, Kassirer seems to have had second thoughts over the Safe affair. He went on to author a book on "corruptive influences" in scientific research titled *On the Take: How Medicine's Complicity with Big Business Can Endanger Your Health.*)

Dr. Safe's own response as to why he didn't tell the *Journal*'s editors about his conflict of interest before the publication of his editorial was nonchalant, at best. Safe told Tye that his views on PCBs and other environmental contaminants were established before he received funding from the Chemical Manufacturers Association. And besides, he added, "There's hardly any life scientist in the country who hasn't had funding from the industry." Nonetheless, Safe said he could see why people would bring it up. He told Tye that he did feel a "little" twinge about not informing the *Journal*, but he added, "it was not much of a twinge."

Dr. Safe's demeanor became much more ruffled and angry in his in-depth interview with Doug Hamilton of PBS's *Frontline* a few months after Tye's *Boston Globe* exposé. Responding to charges of conflict of interest leveled against him by Dr. Fredrick vom Saal, a highly respected developmental biologist concerned with the health effects of industrial chemical pollution, Safe spat, "If Fred vom Saal and his ilk think I lie for industry, I can tell him he's crazy." And he added, "I think it's a McCarthy-like tactic. And it's an outright lie."

Calming down—but with Hamilton still probing—Dr. Safe told the PBS reporter and producer that for approximately four years he didn't "take any personal money." Elaborating, he said, "I get a good salary from Texas A&M University and I'm not profiting from these grants at all." But when Hamilton asked Safe if he received and kept honorariums, the answer was in the affirmative. Back on the defensive, the professor said, "over the last three or more years any consulting money I get I give to the university, with one exception. I'm involved with a drug company in developing a new anti-cancer drug, and I'm on their scientific advisory board and I get paid a little money from them."

The relentless and punctilious reporter, referring to his research on Dr. Safe's finances, pointed out that according to his figures, Safe was being personally paid more than $100,000 per year by the unnamed drug company. The professor didn't dispute the overall sum, but he told

Hamilton the extra six-figure paycheck from the drug company "doesn't put any pressure on me at all."

Later, in a paper published in an academic journal, Dr. Safe defended himself against the conflict-of-interest charges with the seeming mind-set of a defense attorney in a personal injury action. "At the time, based on the then-current guidelines, which asked for current support, I had not declared my previous grant support from the CMA [Chemical Manufacturers Association]." And then he added, in a tepid mea culpa, that "in retrospect, I agree that full disclosure, even of potential conflicts, is the best course and I should have been more perceptive of this issue."

But the apparent damage Dr. Safe had done with his "chemophobia" editorial was not to be easily rectified. Dr. Kirsten Moysich, the first researcher to carefully document the link between breast cancer, PCBs, and women who were genetically susceptible, was forced to write in *The Ribbon*, a breast cancer awareness publication, that positions such as Safe's had created distorted reporting in the media. "Some of these media reports . . . discredited members of the environmental and breast cancer advocacy community, as well as the scientific community, who are not entirely convinced that chemicals like PCBs and DDT do not play a role in the development of breast cancer."

Worse, Dr. Safe had seemingly misrepresented the views of one of his own research colleagues regarding the relationship between PCBs and breast cancer. Dr. Mary Wolff, a coauthor of the Hunter Study, on which Safe based his editorial, was mentioned by Safe as a prominent expert who didn't believe there was any relationship between breast cancer and PCBs. Yet when questioned by the media on the Hunter Study and Safe's editorial, Dr. Wolff told the *New York Times* that it was premature to abandon the breast cancer/PCB link. "It may be important in some groups of women and it may be not only how high the levels are but the time of life in which they occur. Maybe it's even different for different kinds of breast cancer, like premenopausal and postmenopausal."

Dr. Safe's detractors, of which there are many in the medical and scientific communities, scoff at his saying that industry funding of lab science has no impact. Dr. Fred vom Saal told PBS's Doug Hamilton, "So let's say you get a million dollars and it comes into your lab and you set up an infrastructure based on that million dollars. And people's jobs depend on you. There are pressures associated with maintaining the funding and keeping that environment going, and anybody who would claim that getting money has no influence on your behavior I just think is not making a credible argument."

In fact, Dr. Safe's statements on *Frontline* seemed to contradict his own views and bolster Dr. vom Saal's criticism that scientists like Safe are all too aware of the need to get funding for their research. Safe told Doug Hamilton that the rat race for scientific funding could be described as brutally competitive. "We all jump on bandwagons, myself included. Particularly if you're a scientist, because you're always looking for funding."

Responding to Dr. Safe's accusations that Dr. vom Saal was calling him a liar, vom Saal explained that receiving funding from industry doesn't necessarily cause a scientist to lie. "I am not suggesting that anybody is overtly lying. You don't need to do that in science. It is very easy for someone who understands the way a system works to set up an experiment to find exactly what you want to find."

CHAPTER 9

Killer Whales and the Weight

The trail of Homo sapiens, serial killer of the biosphere,
reaches to the farthest corners of the world.

E. O. WILSON, *The Future of Life*

According to Harvard's renowned zoologist E. O. Wilson, since the last ice age there have been more species of plants and animals living on Earth than at any time since the planet was formed four billion or so years ago. If Wilson is correct, then natural climate change—even with the mass extinctions that invariably follow—may have actually somehow facilitated the tremendous biodiversity that Earth has recently enjoyed. At the least, climate change could not have hurt the fundamental ability of life to thrive on the planet too much, regardless of how fatal a glacial ice sheet could be for swaths of flora and fauna.

Now scientists like E. O. Wilson believe Earth is facing mass extinctions once again, not from natural cycles, but from the results of human activities—perhaps the most frightening example being global warming. However, *if* human-induced global warming stays within the range of Earth's historical climatic fluctuations—though mass extinctions will inevitably occur—this shouldn't pose a long-term threat to the viability of the biosphere, *if* past geophysical behavior holds true. Admittedly the above are terrifyingly important *if*s. However, if we get beyond them, then perhaps the greatest danger to biodiversity may come from the tens of thousands of industrial chemicals—some virtually indestructible—that we've allowed to soak into the fragile skin of our entire planet.

The most damaging, the most widely dispersed and longest-lasting industrial chemical group, in the minds of many scientists, are the 209 varieties of PCBs. We've seen that the weight of evidence makes the "magic fluid" a serious, sometimes mortal threat to the health of tens of millions of human beings worldwide. But that is an anthropocentric perspective. How PCBs wreak their havoc on the rest of Earth's living organisms is an unfolding story, and given the current flux in scientific thinking, perhaps it's best to use a real-life parable to describe the danger.

There has never been a documented case of an *Orcinus orca*—a killer whale—killing a human being in the wild, though they could do so with ease and they might even enjoy the supposedly sweet taste of human flesh over their usual fare of seals, large fish, and other cetaceans. Unfortunately for killer whales, their taboo on killing us has not been reciprocated, except for a few tribes of indigenous people in the Canadian Pacific who fear retribution from these formidable predators.

The first recorded commercial hunting of killer whales, or orcas, dates back to the 1700s in Japan. In the 1800s, Yankee whalers—the world's most efficient fleet of killers in their time—would avoid orcas since they had comparatively little blubber, were hard to process, and the larger right and sperm whales were plentiful. Many a whaling captain, though, was known to have his crews harpoon and kill orcas to keep his men sharp when there was a dearth of bigger quarry.

By the 1920s, "factory ships" and mechanization made the killer whales commercially viable targets, and thousands were slaughtered for meat and oil. In Japan, killer-whale flesh was—and still is—prized for its flavor. Others, such as the Norwegians and Soviets, ground their orcas up and used their meat for fertilizer. Now most civilized countries protect killer whales from commercial slaughter.

Orcas have been protected in the United States since 1972 and in Canada from 1994. However, there is disagreement among the members of the International Whaling Commission as to the status of killer whales. The commission has banned the killing of orcas by so-called factory ships, but they can be killed legally by other, unspecified means. In fact, the only codified international regulations enacted covering killer whales extend solely to the trade in their body parts.

For fishermen, killer whales have always been problematic. Longline losses to a single pod of orcas can supposedly range from 50 to 100 percent, and because mature killer whales can consume as much as three hundred pounds of fresh fish in a day, they are indeed troublesome rivals for commercial fishing interests—so troublesome that one salmon fisherman in the Pacific Northwest reportedly said, "It was always possible to know when the killer whales were coming through the gillnet fleet in the Haro Strait by listening to the sounds of gunfire."

But orcas are far more valuable alive than dead. A live killer whale can now fetch more than $1 million on the international market driven by aquariums and marine parks. Of the 250 or so killer whales caught in the Pacific Northwest for aquarium programs, more than a dozen died during capture. And in line with the story of commercial fisherman shooting orcas, 25 percent of all the killer whales captured bore bullet wound scars.

While the commercial fishermen of the Northwest may hate killer whales, the citizens and scientists of the region revere and study them avidly. Pacific Northwest killer whales are exciting to study for many reasons. First—and few scientists would admit this as their motivation—the waters of Puget Sound are beautiful. Snow-capped peaks of the Cascade and Olympic mountain ranges preside majestically over forests that reach right down to the shores of the Sound. More to the point, the waters are mostly calm, given their distance from the open Pacific, which makes for easy observation. And besides being flawlessly adapted creatures of grace and high intelligence, killer whales are naturally suited for research because of their anatomy.

Since the early 1970s, scientists have been photographing and cataloging the dorsal fins of orcas on Puget Sound. Dorsal fins are like finger prints—or, perhaps more aptly, faces. The shape, pigmentation, and scars of each are unique. Killer whales live in stable matriarchal pods that range in size from fewer than a dozen up to forty members. Consequently, by identifying a single dorsal fin within their catalog of photographs, researchers can usually tell exactly which pod they are observing.

Also intriguing for the scientists are the biological parallels that orcas share with humans. Killer whales are susceptible to many diseases that we suffer from, such as Hodgkin's disease, atherosclerosis, cancer, and even gum and dental problems similar to those that elderly humans endure. While these multi-ton giants live in an entirely different environment from us, if you average the male and female life spans they are similar to modern human longevity, with wild males having a life span of fifty to sixty years, and females, who bear young every three to five years, potentially capable of living for eighty to ninety years.

But the Puget Sound resident killer whales—the most celebrated wild marine mammals in the world—are dying. They may be extinct within as few as thirty years if scientific projections are correct. During one four-year span in the mid-1990s their population dropped from ninety-six to eighty-three whales. By 2001 they were down to seventy-eight. And what is killing them and their transient kissing cousins, orcas from the coasts of Oregon and California, is a complex, intriguing, and terrible riddle.

One theory is that the Puget Sound orcas are starving because of the reduced salmon runs caused by habitat destruction due to the urban growth of Seattle, Tacoma, and Olympia as well as Vancouver, British Columbia, to the north. But this doesn't seem to be a productive theory since the transient pods from the south feed on seals almost exclusively and they appear to be dying off as well. And further support for the implausibility of the decline in salmon runs causing the orcas' slide to extinction is the fact that killer whales are consummate hunters. Some experts believe that if prey suddenly became scarce, they would move to areas where salmon were much more plentiful, such as up the coast to British Columbia or even southern Alaska, or even more likely, they would hunt new prey.

There was a theory that the killer whales were being stressed to death by commercial whale-watching outfits that ply the waters around Puget Sound hauling enthralled tourists to gape at the resident pods. In fact, two scientific studies of resident killer whale behavior did indicate that the whale-watching boats were indeed stressing out the whales. Using the number of spouts as a measure of their respiration, the scientists concluded that pods with whale-watching boats nearby had higher respiration rates versus those pods that were hunting without being followed by tourists—and the higher rates supposedly indicated that the animals were being stressed. But this theory of extinction by tourist stress wasn't compelling for many since captive killer whales, while they do die during capture, don't seem to die from stress once in captivity, even though the strain of training, daily shows, and a wholly unnatural habitat must be extraordinarily difficult for the animals.

A few researchers even thought that other researchers were killing off the killer whales. Because individual orcas in Puget Sound are well identified through their dorsal fins, and since they are relatively long-lived creatures, they make ideal study subjects for measuring biomagnification of toxic contaminants in the waters of Puget Sound—something like giant marine canaries. For this reason, and to monitor the health of the pods themselves, researchers take "biopsy samples" from the killer whales by shooting a small dart with a sharp-edged cylinder into their skin.

Using an air gun or a crossbow, the scientists fire as the killer whale surfaces. The dart goes in less than an inch but can supply a wealth of data. Skin samples are used for genetic studies that can pin down the geographic and taxonomic lineage of individual whales with precision. But it is the blubber that yields the most important data. It can be analyzed for nutritional profiles, physiology, and—most importantly—for contaminant levels.

While biopsy sampling is vital for understanding killer whale health, it was associated with an unexpected outcome for some of the resident Puget Sound orcas. Eight out of eighteen whales that had biopsy samples taken in one study conducted between 1995 and 1997 were thought to have died—an extremely high mortality rate. However, there is much doubt as to a cause-and-effect relationship. Killer whales survive high-powered rifle wounds as documented by the large numbers of captured animals with scars. Thus a one-inch wound on their skin could hardly be life threatening, it would seem.

No, other things had to be killing the orcas.

Dr. Peter Ross is probably the world's foremost expert on the health of Puget Sound killer whales. He divides his time between lectures to rapt students of marine mammal biology, field work, and laboratory analysis. Just a few minutes from his modern offices at the Institute for Ocean Sciences overlooking the Strait of Georgia in British Columbia, Ross can be motoring along with pods of orcas—observing, photographing, and "biopsy sampling" them. But his scientific expertise doesn't stop with Puget Sound killer whales. Ross is also known for his innovative research with Baltic seals and their widespread immunological problems, and those studies, conducted in the early 1990s, had critical importance for the Puget Sound killer whales and their looming extinction.

Dr. Ross determined that virus plagues, which killed thousands of seals in European waters in the 1980s, were probably caused by suppression of their immune systems, most likely by industrial chemicals. With the huge amounts and many types of biocidal industrial chemicals that have been dispersed in Earth's biosphere, it was impossible for Ross to unequivocally state which particular chemical caused the deaths of the more than twenty thousand seals. Given this type of scientific data-acquisition nightmare, experienced researchers like Ross fall back to a reasonable position for drawing conclusions: they use the weight-of-evidence approach.

For Dr. Ross, the weight of evidence was strong that there was a star chemical villain. He believed that PCBs, perhaps sometimes working synergistically with other trace industrial pollutants, were the lethal causative agent, not only for the European seal die-offs, but the plunging numbers of Puget Sound orcas.

In fact, the European seals' body burdens of PCBs were only moderate compared to what Ross found in the resident and transient Puget Sound killer whales. Of the three discrete populations of whales in

Puget Sound, the northern residents had PCB levels of thirty-seven parts per million, about equal to the sick European seal populations Ross had studied, but still ranking them well up on the "most polluted" marine mammal list. The southern residents were worse, with PCB contamination rates higher than even the unhealthy Beluga whales of the St. Lawrence in Canada—a species on the very verge of extinction. These southern resident killer whales averaged 146 parts per million of PCBs. And the transient orcas whose blubber Ross analyzed were higher yet, at 251 parts per million of PCBs—making them some of the most PCB-contaminated vertebrates on the planet.

What may have accounted for the extremely high contamination in the transient killer whales was their diet. It was comprised almost exclusively of seals—seals heavily contaminated themselves with PCBs due to biomagnification, whereas the resident whales were known to only eat salmon. (Salmon contamination levels of PCBs are usually far lower than levels that burden marine mammals like seals.)

In any event, the two indisputable facts—that killer whales feeding in Puget Sound were highly polluted with PCBs and they were also dying off—raised the question: if PCBs were the deadly agent, what might be the biochemical mechanisms of their lethality?

First and foremost for mature orcas, it appeared to researchers like Dr. Ross that it was the ability of PCBs to compromise the immune systems of the animal. Ross knew from his extensive research into seal mortality in the Baltic and from other well-executed dolphin studies in the Gulf of Mexico, that PCBs are powerful suppressors of marine mammals' ability to fight infection. However, testing the immune suppression theory for animals like wild killer whales is all but impossible. When they die their carcasses are rarely found. Consequently, even getting a chance to forensically pinpoint what killed any given orca whale is an exceedingly rare event.

But just such an occasion presented itself in the spring of 2001 with "Everett," when he washed up dead on a public beach on Vancouver Island. Known to scientists as J18, the twenty-something male was a member of one of the most studied pods of killer whales on Puget Sound. On being examined, he was found to have an acute systemic infection caused by what was usually a harmless bacterium. Scientists who performed the postmortem felt that Everett probably died from a stomach abscess, a disease that shouldn't result in death for a male killer whale in his prime years, as Everett was—unless, that is, his immune system was fatally compromised.

Everett was also emaciated. This too would tend to confirm the weight of evidence of PCB-related immune suppression. In the case of Everett, the theory would go something like this: Since PCBs accumulate in fatty tissue, if prey becomes scarce for a killer whale, as can happen between salmon runs in Puget Sound, then the animal starts burning fat to make up for the caloric shortfall. As a result, the killer whale unlocks and metabolizes PCBs as he burns up his fat reserves. By reintroducing PCBs into his circulatory system and organs, where they are known to disrupt the manufacture and deployment of critical immune agents such as T cells and other killer cells that attack and destroy viruses and bacteria, the killer whale compromises his own immune system.

Data supporting the link between PCBs and immune suppression in marine mammals is strong. Besides Ross's research with Baltic seals, Dr. Garet Lahvis, a skilled and frequently cited researcher in marine toxicology now working at Oregon Health and Science University, established in the mid-1990s that PCBs caused reduced immune system response in bottlenose dolphins from the Gulf of Mexico. Studies of dead beluga whales found in the St. Lawrence River also were very suggestive of PCB immunotoxicity, given the incidence and severity of bacterial skin lesions that researchers found on the dead animals combined with their heavy loads of the more toxic varieties of PCBs in their blubber.

Another potentially lethal relationship that PCBs have with immune suppression in marine mammals involves the manufacture and transportation of vitamin A, a vitamin critical to the activation and arming of the vertebrate immune system. (Recall Dr. Frederick Plapp's research on vitamin A and human immunological problems, discussed in chapter 7.) When orcas—just like humans—are exposed to PCBs, the PCBs can poison the processes involved in making vitamin A by chemically corrupting the carrier protein transferrin—a protein that distributes vitamin A and other essential thyroid hormones.

The second deadly threat that PCBs present to killer whales involves reproduction, the primary factor in species' existence. Not only are members of the Puget Sound southern resident pods apparently dying prematurely, but the pods are not reproducing enough calves to repopulate their groups. The precise cause of their possible collective reproductive failure is a mystery and may remain so given the intractable problem of doing experiments with wild killer whales. Nonetheless, there is a persuasive body of data that PCBs are to blame—and again vitamin A metabolism might play a villain's role.

Soon after investigators started studying the contaminant loads of the Puget Sound killer whales, they discovered an intriguing anomaly. Female killer whales of calf-bearing years usually had low levels of PCBs in their blubber. Yet prepubescent female killer whales and older females beyond reproductive age had the same exceptionally high body burdens of PCBs as their male counterparts carried.

There was only one good explanation for this phenomenon. The bearing females were passing along huge loads of PCBs to their calves via lactation. Killer whale milk is tremendously rich in lipids (fats), since the calves need a quick, thick layer of fat to provide insulation against the frigid waters in which they are born. Since PCBs are lipophilic (fat-loving), the female killer whales dose their calves with large amounts of polychlorinated biphenyl along with their life-giving milk. By the end of lactation, a female killer whale might have transferred up to between 75 and 95 percent of her body burden of PCBs to her calf. (Part of the killer whale contamination transfer involves the birth order of the calves, with the first-born calves getting by far the largest hits of PCBs.)

Again, the real-life significance of PCB contamination via lactation for wild killer whales cannot be studied directly since the animals are free-ranging. However, Dr. Ross and other researchers have analyzed the reproductive impact of disruption or depletion of the vitamin A supplies on captive harbor seals. The research using seals as marine mammal proxies for killer whales unmistakably points toward vitamin A disruption by PCBs as a central factor that should be considered in the apparently high calf mortality in the Puget Sound killer whale populations.

Vitamin A is essential for growth and development in all mammals, but young marine mammals especially need a good ration of vitamin A to develop normally in their demanding marine habitat. However, just when their requirements for vitamin A are greatest, PCBs zap not only the transfer of vitamin A from mother to calf in milk, but they also upset the calves' ability to produce and utilize vitamin A for themselves.

The possibly fatal relationship between PCBs, vitamin A, and killer whale reproduction may in fact be even more oblique and insidious than what we've just seen. Evidently, a significant number of Puget Sound female killer whales are either not bearing young or never have. Since vitamin A is vital for cellular differentiation during the earliest stages of development for killer whales, if vitamin A metabolism is disrupted enough, then there is a substantial chance of abnormal embryonic development and consequent mortality due to spontaneous abortion of the killer whale fetus. So PCBs, by disrupting vitamin A's reproductive functions, may be preventing some killer whales from bearing calves at all. But as usual, the direct evidence of this type of PCB-induced mortality for wild killer whale populations is probably impossible to garner, even though scientists came close in the winter of 2002.

At that time a healthy female killer whale washed ashore on the Dungeness Spit, a small peninsula that juts into the Straits of Juan de Fuca at the entrance to Puget Sound like a large, sandy, flexed arm. She had last been observed in 1996, well down the Oregon coast, and her carcass appeared to be that of a healthy twenty-two-year-old, sleek and well fed. But when a researcher with the National Marine Fisheries Service made a routine PCB analysis of her fatty tissue, the result was too high for their equipment to read. "She basically knocked our instruments off the scale," the investigator told the media. After recalibrating, it was determined that the killer whale had approximately 1,000 parts per million of PCBs in her blubber.

It was the highest PCB reading ever found in a female killer whale of reproductive age. Since female killer whales reach sexual maturity at about the same age as human females, and given the usually able and active male response to this coming of age, female orcas bear calves quickly upon reaching puberty. That this female had such a huge body burden of PCBs could only mean that she had never calved, for if she had, she would have off-loaded a good proportion of her PCBs to her calf via lactation.

The possible causes for her not having calved varied. Either there were no males who could get her pregnant—an unlikely circumstance—or she was somehow rendered infertile, or perhaps she bore calves that died at birth and therefore never suckled, or suckled for too short a time to reduce her PCB burden.

It was mysterious, her death from unknown causes, her apparent barrenness, and her astonishing PCB contamination. But perhaps the most basic question was unasked in the extensive media coverage of the whale's demise: just where were the PCBs coming from that contaminated her and other Puget Sound orcas?

Like us, killer whales get the great majority of their PCBs from what they eat. Resident Puget Sound orcas feed mostly on salmon, hunting them in gregarious packs. In the case of the dead transient female, she was a member of a population of killer whales that feed mainly on salmon-eating seals, silently stalking them alone. But regardless of the differences in feeding techniques, both groups of killer whales have salmon as their direct or indirect food source. Thus the question becomes, where did the salmon pick up *their* PCBs?

At this point, scientists have no definite answers. Chinook salmon that return to the Puget Sound region to spawn are found to have an average of two parts per million of PCBs by body weight, a high figure to be sure. Puget Sound has many sources of PCB pollution, such as the large naval base in Bremerton, the industrial areas around the Duwamish River and Elliott Bay near Seattle, and Canadian sewage treatment plants, to name just a few. All have elevated levels of PCBs in surrounding sediments and around their outfalls. Certainly the salmon were picking up some of their body burdens from these areas as the PCBs flowed or leached into Puget Sound. However, salmon only spend a relatively short period of their life cycle near their ancestral spawning grounds. Most of their lives are believed to be spent swimming in the northern mid-Pacific. So then the question in turn becomes, where were those open-ocean PCBs coming from?

The most likely source is Asia, through airborne transportation of the PCBs. This theory is backed by studies of bald eagles, sea otters, and black-footed albatrosses found on remote islands in the mid-Pacific and the Aleutian chain. All were discovered to have comparatively high levels of PCB contamination despite their distance far from known industrial sources of pollution.

PCBs may threaten far more than the Puget Sound killer whales with extinction. While there is much controversy about it, PCBs might jeopardize entire marine ecosystems. In the early 1990s, researchers who had been studying northern Pacific sea otters noticed a steep decline in their numbers, so steep that it became apparent they were actually being wiped out. By 2003, 96 percent of the sea otters in the Aleutians had disappeared. (Eight out of ten of all the world's sea otters lived in the Aleutians prior to their virtual extirpation.) After analyzing the sea otter catastrophe, scientists discovered that they were being eaten by killer whales.

This was highly puzzling behavior for the orcas. They had lived in harmony with sea otters on the Alaskan coast for millennia, even though their peaceful relationship with otters certainly wasn't pure beneficence on the part of the killer whales. Sea otters are small, relatively lean prey that can hardly provide more than a snack for a mature orca. Traditionally, Alaskan orcas fed on much bigger marine mammals with lots of caloric content—fatty seals and blubbery sea lions.

The reason for the dietary change of the killer whales soon became evident to the researchers. The orcas had started eating sea otters because their usual prey was gone and the otters were simply targets of opportunity for the hungry orcas. The stellar sea lions—the preferred food of the northern killer whale—had almost become extinct in the Aleutians by the early nineties, with only dozens left in areas where there had once been thousands. Harbor seals and fur seals had declined precipitously during this same period as well. The killer whales, highly intelligent animals, were forced to improvise and change their diet.

What would cause such a steep decline in the population of seals and sea lions in the areas surrounding the Bering Sea? Some experts believe that PCBs—both airborne and from dumping by military installations that dot the Aleutians—could well have played a key part in the virtual multispecies extinctions given the chemicals' toxic effects on seal and sea lion immune systems, as well as the disruptive impact that PCBs are known to have on their reproduction.

Whenever top predators are extirpated from an ecosystem, delicate natural checks and balances are almost always destroyed as well. So it apparently was in the Aleutians. With the decimation of the sea otters by orcas, the sea urchin population exploded, beginning what researchers call a "trophic cascade." The sea floors became covered with the spiny creatures, with scientist divers reporting dozens per square foot in places.

Although not usually voracious feeders, the hordes of urchins—without sea otters to prey on them—started wiping out the kelp beds, their favorite food. And within a few short years the kelp beds were mostly destroyed. This created an ecological catastrophe, since kelp beds are the main nurseries and habitats of much of the marine flora and fauna of the Aleutians. In approximately five years, a sort of marine equivalent of a gigantic tropical rain forest had been laid bare.

The Alaskan orcas turned otter killers were also found to have high loads of PCBs themselves and appeared to be in decline. A well-monitored killer whale named Eyak washed up dead in July of 2000 near Cordova, Alaska, with 470 parts per million of PCBs in his tissues—very high readings indeed. Another transient male was biosampled in the Gulf of Alaska and found to have the highest PCB burden ever measured in a living northern-dwelling orca—651 parts per million. (It was also noted that his dorsal fin was drooping, an almost sure sign of ill health, if not approaching death, in killer whales.)

Yet in reality, nobody knows with certainty just how healthy or unhealthy the Alaskan and Pacific Northwest orcas are as populations or individually. There just isn't enough information. Evaluating a single sample of blubber for PCBs can cost thousands of dollars. The testing equipment used by scientists consists mostly of gas chromatographs along with electron capture detectors, fragile instruments that require a high degree of understanding and skill to operate, not to mention the laborious and exacting process of preparing tissue samples for analysis. And estimates of the testing costs don't usually take into consideration the funds needed for boats, crews, and darting equipment and the fuel costs involved in finding and taking biopsy samples from wild killer whales.

The result is that despite the keen popular interest in these mythical sea creatures, even a rich country like the United States hasn't underwritten enough research to truly determine the extent of PCB contamination from which orcas suffer—and if that contamination is responsible for their decline and possible extinction. Currently, it is all a weight-of-evidence approach, with the good possibility that by the time there is absolute scientific proof of PCB lethality for orcas, they will be long gone.

A Lethal Erosion of the Biosphere

Through their suffering the frogs give us a sharp,
clear warning of a lethal erosion of the biosphere.

E. O. WILSON, *The Future of Life*

It is odd how a bundle of paper with traces of ink can direct a person's existence. After receiving his degree in molecular zoology, Arjan Palstra was trying to decide what track his life should take. As a kind of entrée into the world of scientific research, Palstra's uncle, a well-known Dutch biologist at Wageningen University, invited him to help study small colorful fish called "barbs" in Lake Tana in Ethiopia. While thinking over the offer, Palstra visited a favorite bookstore and happened on Tijs Goldschmidt's unheralded literary science classic *Darwin's Dreampond*, the story of the Goldschmidts' experiences researching fish extinction in Lake Victoria.

Palstra read the book, went to Africa, did fish research, and was hooked—so to speak. On returning to the Netherlands, Palstra applied for graduate school at the distinguished fish biology department at Leiden University. He was accepted and promptly fell in love—with one of the ugliest and most mysterious creatures wrought by the creative hands of Mother Nature, *Anguilla anguilla*, the common European eel. But sadly, it was to be a tragic love affair from the start. Of the tens of billions of eels that had lived in the rivers, streams, ponds, and lakes of Europe for millennia, approximately 99 percent had disappeared.

While the squeamish may be repelled by eels, scientists like Palstra can be easily smitten by their biology and design—for eels are animals that are precision crafted by evolution to survive and thrive in perhaps more different environments than any other species of land or marine animal. At different times in their life cycles eels live in everything from muddy lake sediments with little oxygen to cascading streams and rivers, to deep ocean realms.

It is believed that the strange life of the eel—European and American—begins in the Sargasso Sea in the midwestern Atlantic. Although nobody has seen or found adult eels in the Sargasso Sea, it is likely they spawn there since their larvae—diaphanous, tiny micro-eels called *Leptocephalus*—are found in the nearby Gulf Stream, which apparently acts like a conveyor belt transporting the larvae to the east coast of America and to northern Europe.

European eels are catadromous. This means that, unlike their anadromous fish colleagues such as salmon and steelhead trout, they spawn in the ocean but live in freshwater as adults. When the eels' biological alarm clocks go off and they are drawn toward spawning in the Sargasso Sea, the urge is so powerful that if they must, they will squirm and wriggle overland to reach sea-bound streams or rivers.

They also undergo a great metamorphosis. After they have lived on stored fats without feeding for their entire spawning journey, the digestive tracts of the eels dissolve, helping to make their already streamlined bodies become even more hydraulically efficient. Their eyes enlarge and develop special pigments that allow them to see more easily in the dim blue light that deeper oceanic travel requires. Their bodies become silvery, making them less visible to the predators that wait to ambush them on spawning runs thought to be more than four thousand miles in some cases.

(The eels' system of navigating long distances in the ocean is unknown. It is likely that they use Earth's magnetic fields, perhaps like migratory birds do. Researchers at Leiden implanted microchips into the muscles of female eels and monitored them over a period of months. They must have grinned when they considered the study data. The eels oriented themselves in a south-southwest direction each night—the same direction as the Sargasso Sea.)

The biomechanics of how eels actually swim to the Sargasso Sea is, in and of itself, semi-wondrous. Having gorged themselves before migration, they will travel the thousands of miles necessary to reach their spawning grounds on fats stored in their two- to four-foot long bodies. (Remember that they can no longer feed since their digestive tracts have self-destructed.) With their long, tapered bodies and small, close-in fins they are perfectly adapted for long-distance travel. So faultlessly has evolution designed them that they are the most efficient users of energy of any fish studied. Eels' metabolism and utilization of oxygen make them five to six times more efficient at oceanic travel than seagoing trout or salmon. (To get an idea of just how energy efficient the *Anguilla anguilla* is, compare the heat energy stored by eels in their fat with the energy contained in gasoline: if an eel could burn gasoline, then an average-sized eel would get approximately twenty-four *thousand* miles per gallon.)

Any creature that is so superbly adapted for survival is bound to be abundant, and given the tens of billions of eels that swam in waterways all over Europe, they were indeed plentiful, almost obscenely so—that is, until around 1980, when they mysteriously started to vanish.

Not only was it a nearly incomprehensible natural disaster that such an immensely prolific species of fish could virtually disappear, there was a human dimension to the calamity as well. Many Europeans consider eel to be a great delicacy with its firm, sweet, fatty flesh that lends itself to smoking and broiling like no other. At least twenty-five thousand fishermen made their livelihood by catching eels—more fishermen than for any other species of fish or any wild food source in Europe.

No one had been able to ascertain why the eels were disappearing in European waters. There was speculation by researchers that the decline was caused by global warming, a possible scenario being that the Gulf Stream's transporting ability had been dissipated because of increasing ocean temperatures, which in turn led to the eel larvae not being carried back to Europe. But this hypothesis seemed to be in error, since it so happened that at approximately the same time—in 1980—eel populations in Asia started crashing. Their decline, of course, could not be caused by disruptions to the Gulf Stream.

There was also the usual speculation about overfishing, which is usually trotted out when marine biologists don't have any good ideas for why a fish species is dying out. But there wasn't a major increase in the amount of eels caught in the mid- to late 1970s, nor were there greater numbers of fishermen plying their trade. Furthermore, restoration and management programs that included fish ladders for the eels, as well as quotas for fishermen, didn't produce any increase in the numbers of returning eels.

Another theory was that the eels weren't getting enough to eat before going on their migrations. They weren't storing enough fat because of diminished food sources, resulting in the eels not having enough energy reserves to reach the Sargasso Sea. But when Leiden University fish biologists researched this possibility by testing eels for their metabolic rates, they discovered that eels used only 40 percent of their stored fat on their multi-thousand-mile migration. (The excess fat reserves would be used by the eel's gonads for reproduction.) So, it was unlikely that lack of food was pushing the European eel toward extinction.

It had to be something else that had caused them—billions of them—to disappear.

With all the eel researchers in Europe flummoxed, doctoral candidate Arjan Palstra, along with his mentors at Leiden, decided to do some scientific sleuthing around 2003. Three questions needed to be answered. First, what was it that could affect eels all over the globe? Secondly, was there anything in the biology of eels that set them apart from other fish, and made them susceptible to extinction? And finally, might their habitat have a bearing on their decline?

The answer to the first question was easy. Yes, there were agents that contaminated the entire planet and affected fish and many other things—industrial chemical pollutants. And of these toxic substances, what might be the most prevalent and poisonous to marine vertebrates like eels that migrated through, and lived in, waters of industrialized countries? That would be PCBs.

As to the second question, yes, there were things about eels that made them unique. From a physiological standpoint they are an extremely fatty fish. Adult eels prior to migration build up body fats to the point where one-third of the weight of a healthy eel is comprised of fat. Another factor contributing to the eel's susceptibility to contamination by industrial contaminants is that eels are a species of fish with relatively long life spans. Females can live up to twenty years and their longevity allows them to store up toxins like PCBs. So was there a global contaminant that was ubiquitous in large amounts that just loved to reside in fat? Sure. That, too, would be PCBs.

What about the habitat and behavior of eels that might make them especially vulnerable to man-made pollutants? Again they were unique. Eels spend much of their time residing in sediments—muddy areas of lakes, streams, and rivers. They were the only major food fish that lived in such habitats and also migrated into the open ocean. Why was the sedimentary lifestyle of the eel a problem? Because a group of toxic industrial chemicals called organochlorines (of which PCBs were and still are usually the most common) pollute sediments. In fact, second only to fat, PCBs prefer sediments—oozy, muddy places—as a residence.

What Arjan Palstra wanted to find out was whether the body burdens of PCBs that the eels carried could impair or actually block reproduction to the point where *Anguilla anguilla* could be wiped out. The idea went like this: Female eels burned fat from their bodies during their long migration runs. Once their fats, often loaded with a particularly toxic variety of PCBs, were mobilized into the circulatory system of the eels, some of the PCBs—perhaps a majority—went to the gonads of the female. There they would be incorporated into energy packets for their embryos.

With some careful and scientifically stylish laboratory experiments with live eels, Arjan Palstra found what nobody else had. The exposure of eel embryos to toxic levels of PCBs would deform the embryos to such an extent that there would be almost zero chance of their survival. Female eels from all over Europe may have been reaching the Sargasso Sea with loads of doomed eggs. As each successive generation of mature eels migrated and died, there would be no replacements.

There was even more that pointed to PCBs as the culprit in the extirpation of the eels. It was also obvious from other studies by European scientists that the eels' immune systems had been compromised. *Anguilla anguilla* suffered from two serious and species-wide diseases: a virus called the EVEX (for Eel Virus European X), which caused hemorrhaging and anemia, and a parasite—a nematode called *Anguillicola crassus*—that lived in the swim bladder of the eel, rasping through tissue and sucking its blood. Both diseases were added stressors that made the long eel migration that much more problematic.

The EVEX virus disease in the eels was pretty clearly related to suppressed immunity due to their contamination with PCBs and other organochlorines. But could PCBs cause the eels to be more susceptible to infestation by a nematode—a worm? The answer was yes, indeed. Dr. Bernd Sures, a German eel researcher, carried out a series of experiments with eels by dosing them with PCB-126—a common type of PCB found in eels and all sorts of other European wildlife. Dr. Sures discovered that when the eels were exposed to the PCBs, what resulted was a disastrous "complete suppression" of their immune systems' defenses against the nematode worm.

So Swann's "magic fluid" did a one–two number on the eels. If they didn't die en route to spawning from viruses or parasites due to PCBs compromising their immune systems, then the PCBs would deform their embryos beyond survivability. Thus PCBs, so long established as ubiquitous pollutants that they had become more or less ignored by a large part of the world's scientific community, were unequivocally and quietly reestablished as chemical super bad guys by Arjan Palstra and his associates at Leiden.

So bad were they that molecules of Swann's "magic fluid," in mere parts per trillion, were probably responsible for the greatest destruction—in terms of marine vertebrate biomass—in the history of the planet.

It was hardly a coincidence that the United States and Canada suffered the same kind of aquatic eco-disaster that the PCBs presented to Europeans—not with eels, but with one of America's biggest freshwater fisheries. Ten years before Arjan Palstra completed his innovative eel investigations, two American scientists, Dr. Phil Cook of the Environmental Protection Agency and Dr. Richard Peterson of the University of Wisconsin, discovered that PCBs and PCB-like dioxin byproducts were probably responsible for the commercial extinction of lake trout in Lake Ontario.

Back in the 1930s and 1940s, Ontario lake trout were caught in huge numbers—some twenty million pounds annually. The top aquatic predator of the Great Lakes, lake trout were not only abundant and delicious, they were beautiful silvery fish that could weigh up to thirty pounds and live for decades—a sort of crowning evolutionary glory for the Great Lakes aquatic ecosystems.

Their steep decline began in the early 1940s and by mid-century the fishery was collapsing, with the lake trout headed for commercial extirpation. As the lake trout population nosedived, the ecology of the Great Lakes—especially Lake Ontario—was thrown into chaos. Alewife and smelt populations—the traditional prey of the lake trout—exploded. And then, with the lowered oxygen levels in parts of the lake (due mainly to phosphorus used in detergents and fertilizer, for which our old friend Theodore Swann was responsible), they washed ashore in huge stinking masses, rendering hundreds of miles of Great Lakes shoreline a fetid mess.

Fisheries biologists blamed overfishing along with an introduced parasite—the lamprey eel—for the annihilation of the lake trout populations and the ensuing eco-disaster. The Canadian and U.S. governments instituted strict catch quotas and initiated aquatic control programs for the lampreys. On the U.S. side of the lake, tributary rivers were outfitted with giant lamprey eel capture nets and creeks were doused with chemicals to kill the lamprey eggs. Between Canada and the United States, tens of millions of dollars were spent to control the lampreys.

And in fact, the measures to save the Ontario lake trout were successfully implemented. The commercial fishermen cooperated and limited their catches, even though it was a serious economic hardship that drove many of them to the brink of financial ruin and beyond. The anti-lamprey efforts were relatively successful as well, with the lampreys at least being held in check and not proliferating.

So the fisheries biologists figured that they had the problem solved, and in the late 1970s there was a series of projects that reintroduced lake trout into Lake Ontario through hatchery-raised fish. Millions of lake trout were released. The federal and state fisheries experts waited expectantly for the lake trout to reproduce and multiply back in their refurbished habitat. The experts waited. And they waited. But the lake trout just couldn't seem to reproduce in Lake Ontario anymore. It was back to theorizing, after twenty years of apparently fruitless and costly public effort.

Enter Dr. Phil Cook. Like Arjan Palstra, Dr. Cook decided to go back and do what had needed to be done thirty years before, when the first alarms regarding the contamination of the Great Lakes by PCBs had been sounded by Wayland Swain and others. Cook tested the eggs of Ontario lake trout with coplanar PCBs—a variety that has similar effects to dioxin (an extremely toxic substance that often is a byproduct of manufacturing PCBs). The results of Cook's experiments were profoundly disturbing. At a mere thirty parts per *trillion* of coplanar PCBs, developing lake trout embryos started to die. At one hundred parts per trillion, Cook found 100 percent mortality in the lake trout embryos due to extreme cranial deformations and massive hemorrhaging.

The significance of Dr. Cook's experiments couldn't have been weightier. After checking decades-old PCB survey data from Lake Ontario, Cook found that ambient levels of PCBs in Ontario during the period when lake trout went commercially extinct averaged more than double the amount necessary to kill lake trout eggs. In other words, throughout Lake Ontario, the levels of the highly toxic type of PCBs and the dioxin cohorts were enough to cause 100 percent mortality in the lake trout eggs. The fish could have been laying their eggs on the moon and the result would have been the same.

Are the eel and lake trout disasters representative of what is happening to other fish? Are PCBs killing off other species of aquatic vertebrates? These seemingly scientific questions turn out to be more answerable in political terms. First, one must realize that scientists like Arjan Palstra and Phil Cook are always strapped for money. Theirs is an ethically motivated labor of love, for the most part.

In the United States, big research grants, the literal coin of the realm, mostly go to scientists studying either human medical problems or military ones. As far as corporate research into fauna and the impacts of industrial contaminants, their grants—which can be quite lucrative comparatively—generally go to researchers who tend toward exoneration of industrial chemical pollution. Exacerbating the political and financial aspects of studying the effects of PCBs on fish (and all animals) is that the right analytical techniques require great skill, time, and expense.

Consequently, the real answer to the question of what PCBs are doing to fish globally is, we don't really know. Fish species and individual fish within a species, like humans, react differently to PCBs depending on their individual genetic makeup. Since so little testing has been done and few—if any—thorough surveys of fish within broad ecological systems were ever completed *prior to* PCBs contaminating the planet, we really don't know how many species of fish have been wiped out. It could be many. It probably is many, if the eel and lake trout calamities are indicative—and there's no reason to believe that they aren't.

While untold species go extinct, slide into oblivion without a whimper, without a sound bite or an inch of newspaper copy, there is some room for optimism albeit small. Fishery biologists and toxicologists, mostly on the east coast of the United States, where pollution is heaviest, are reporting that some fish seem to be able to "adapt" to PCBs and reproduce in reasonable numbers in relatively heavily polluted fresh- and saltwater systems. The adaptation appears to be via normal evolutionary mechanisms—heritable changes in genes that allow fish for "at least three to four generations" to survive PCB exposure, according to Dr. Adria Elskus, a researcher with the U.S. Geological Survey.

It's good news—but still worrisome. The fish that are adapting to PCBs are of smaller varieties that reproduce quickly and in large numbers. While they may survive, they will probably still have relatively high concentrations of PCBs locked in their fat, making them dangerous for predators to feed on with regularity. And more unfortunately, there is no research that indicates that top aquatic predators that bio-accumulate PCBs—such as seals and osprey, or freshwater fish eaters like mink or otters—will genetically adapt to the "magic fluid." And it is these high-level predators that ultimately determine the health and long-term viability of ecosystems.

On June 5, 2006, the *New York Times* published a "cute" article on the travails of two scientists going through airport security with suitcases packed full of live frogs. The frog researchers were coming back from a Panamanian rain forest where they had captured hundreds of endan-gered frogs just days before they knew they'd be wiped out by a virulent skin fungus. It was a planned bit of the type of grandstanding that scien-tists are usually loath to engage in, but it made for great newspaper copy and also allowed the reporter to segue into one of the great scientific disaster mysteries of all time.

Since approximately 1970, entire species of amphibians—frogs and salamanders—have been disappearing. Individual creatures were either dying outright by the tens of millions, perhaps billions, or they were fail-ing to produce offspring. The most mystifying and frightening aspect of the amphibian extinction was that unlike the extinctions of other spe-cies that occur in a geographic area or region, the frogs were dying all over the globe. From South American rain forests to Australian deserts, from mountainous terrain in Europe and North America to marshes in the Near East, frog species began vanishing—in some cases within just months of first being identified by appalled field researchers. It was, and still is, an ultimate biological nightmare—beyond true comprehension for the scientists involved in documenting the amphibian extinction. Making it even more surreal for them was that few nonscientists can really grasp the terrifying implications of the disaster.

For the average citizen in an industrialized nation, frogs are just squishy, slimy creatures that, because of their secretive nocturnal hab-its, are rarely if ever seen except maybe as characters on children's TV cartoons. But biologists know better. They know, to their too often unspoken horror, that amphibians are an extraordinarily important class of animals for Earth's ecology. How important can be straight-forwardly illustrated by a study done of tiny cricket frogs in Iowa in the early years of the amphibian extinction: researchers estimated that a thousand cricket frogs living in a small pond would eat more than four *million* pounds of insects in a single year.

Given the voracious appetites that most frogs and other amphibians have for insects, they are a major contributor to the vertebrate biomass of many terrestrial ecosystems. Dr. Robert Stebbins, as a young profes-sor at UC–Berkeley in the mid-1950s, undertook one of the few investi-gations of how amphibians affect biomass by studying a redwood forest in the hills of the East Bay. Dr. Stebbins, later recognized by many as the world's leading expert on amphibians, discovered that two types of tiny salamanders contributed more weight per acre to the biomass of the forest than any other vertebrates—including moles, squirrels, and all species of birds combined.

Not only in the temperate forests and common marshes are amphib-ians crucial, but they are ecologically pivotal in the foundries of biodi-versity, Earth's tropical rain forests. In fact, amphibians are believed to be *the* critical vertebrate link in the food chain of rain forest ecosystems. Frogs, and toads in particular, not only keep insect populations in check, they themselves provide a large, ready source of protein for higher-order predators: snakes, jungle cats, carnivorous birds, and just about all predatory freshwater tropical fish.

But aside from being perhaps nature's best two-way caloric con-veyor belt, amphibians have a unique role in predicting the health of the planet. "Through their suffering the frogs give us a sharp, clear warning of lethal erosion of the biosphere," writes E. O. Wilson in *The Future of Life*. "We ourselves could not have devised a better early-warning device for general environmental deterioration than a frog."

The reasons for Wilson's view are twofold. Frog populations gener-ally are hypersensitive to minute changes in any ecosystem they inhabit since their larvae and tadpoles are bottom feeders where toxins, like industrial pollutants such as PCBs, are likely to accumulate. And per-haps more importantly, in both adult and larval stages amphibian skin breathes, acting as a device for the exchange of oxygen, thus making it a near-perfect natural apparatus for gathering toxic compounds both from the air and the aquatic environment. As Wilson says, "Frogs are nature's canary in the mine."

Nothing could be more menacing for a miner than when the caged canary dies, could it? Well, actually, yes, there is something worse—when, perhaps, the canaries start dying *everywhere* in the mine. For frogs such a turn of events is no longer metaphoric but a reality, and biologists across the globe are feverishly trying to determine the cause of their pending extinction. While the theories are numerous, the most common actual cause of death for individual frogs is believed to be from infection, with the leading killer being a fungus, specifically, the chytrid fungus.

What makes the chytrid fungus the focus of so much interest is that it attacks almost all species of frogs regardless of habitat—be it desert, marsh, or mountainside. And chytrid fungus is found almost everywhere on the planet. It has been linked to the decimation of frog populations from Europe to Africa, from North America to Central and South America, and in Australia as well.

In fact, Australia is the location of the discovery of chytrid, where it "appeared" in 1993. But scientists who began to investigate the disease quickly realized that it had been present in Australia since at least 1978, and since it was soon found everywhere on the planet save Asia, they came to believe that chytrid may have been present in the environment for many years—maybe even millions.

Although investigators quickly established that chytrid fungus caused changes in the skin of infected frogs, the exact way it killed them was a mystery. Since frogs breathe through their skin, one reasonable idea was that the fungus somehow rendered the infected frogs' skin unable to absorb oxygen and they suffocated. Another theory for chytrid's lethality was that it was known to produce minute amounts of a deadly natural toxin, and it was actually the toxin, not the fungus, that was killing the frogs. But nobody knew. (And as these words are written, researchers still don't know the exact means by which chytrid kills frogs, if in fact it does.)

Making things even more of a mystery, and a real-life horror story, it was found that although many species of frogs would become extinct after being infected, not all would succumb to chytrid. Some species of frogs, even though they were exposed to the infectious spores daily, did not contract the disease. Yet another species in the same general habitat would be wiped out by the fungus in only a number of days. In other instances, a species of frog might decline precipitously by 70 percent or even more—only to have its population stabilize after the initial decimation. The diverse responses to chytrid indicated that there probably were major genetic differences in amphibian immune reactions to the fungus from species to species, and even between individual frogs within the same species.

But chytrid fungus was not the only infection that was killing off frogs in large numbers. Dr. Stebbins also found that another bacterial disease appeared to be annihilating frogs in some parts of the Americas. It was called "red leg disease" because it infected the tissue on the underside of the legs of frogs with a reddish inflammation. The bacterium that caused red leg disease was common throughout freshwater systems in North America. For perhaps hundreds of thousands of years, frogs had thrived living with it. Now, all of a sudden, the bacterium had become lethal. Field researchers also discovered that besides frogs being killed off from bacterial infections like red leg, a variety of salamanders was also succumbing to viral infections in North America, viruses that had apparently never been a problem previously.

It seemed obvious to Dr. Stebbins that something—some toxic agent—was compromising the immune systems of frogs all across the planet, rendering them and other amphibians unable to resist infections that for perhaps millions of years had not been lethal to them. So, what agent could potentially compromise the immune systems of frogs across the planet with a sort of amphibian version of AIDS? What might disrupt their endocrine systems to the point where they weren't able to produce enough of the right hormones for their immune systems to combat an infection? And what toxic agent was ubiquitous by the 1970s—and toxic in minute amounts—when the frogs started succumbing en masse to infectious diseases?

Actually there were two prominent worldwide agents that could fit the bill almost perfectly: a DDT metabolite named DDE and . . . PCBs. So which one was it—if it was indeed either of them?

Once again, as these words are written, there is no answer. First, because there are no published studies comparing DDT metabolites and PCBs' effects on the abilities of amphibian immune systems to resist the chytrid fungus or red leg disease. It could be either one, or both, or both in combination with the hundreds of other industrial chemicals that have come to contaminate every cranny of the biosphere. It certainly is possible that PCBs are acting synergistically with some of these, but since global levels of PCBs reached their zenith just when the amphibian declines began, PCBs could well be the sole chemical assassin.

However, for the sake of argument, let us assume that PCBs are interacting with some other industrial chemical and that together they become lethal to amphibians, whereas, individually each separate chemical agent would not be. Even in that situation PCBs would present the greatest biological threat. This is because there are generally more PCBs circulating in the biosphere than any other industrial chemical contaminant, and if you somehow remove the PCBs, or reduce their levels in the environment below some unknown threshold, you render them and the other toxins benign by obviating their synergistic capabilities. This is the "kingpin" theory of ecotoxicology. Remove the kingpin contaminant and you end the threat—at least hypothetically.

Another link between PCBs and global amphibian extinction could be the impact that polychlorinated biphenyls have on amphibian reproductive cycles. Frogs undergo tremendous physiological changes when they metamorphose from tadpoles to adult frogs. The entire process of the near-total reorganization of the frog's body is driven by hormones (almost exactly the same hormones that the human endocrine system deploys, it should be noted).

Because their physiology undergoes such a profound transformation, the hormonal storms that these changes require are some of the strongest in the animal kingdom. Consequently, since frogs have a unique necessity for producing and processing large amounts of hormonal signals, they are in theory probably the most susceptible vertebrates on the planet to the threats of hormonal mimics such as PCBs.

The hibernation process that many frogs undergo puts them at special peril from fat-loving contaminants like PCBs. According to studies cited by Dr. Robert Stebbins, frogs and toads will use up a large amount (50 percent or more) of their stored fat during hibernation. When they emerge from their retreats and burrows, they are always in a state of comparatively great physiological stress due to their depleted energy reserves. This high-stress state also coincides with the point where PCBs are most likely to be circulating in their systems, since PCBs are known to be released when vertebrates metabolize their fat assets. This in turn makes them "particularly susceptible to the impact of internally released chemical contaminants," according to Stebbins.

Perhaps the most important way that PCBs may contribute to, or even be the chief cause of, the amphibian decline is their potential effect on the breeding cycle of frogs—which usually occurs soon after the end of hibernation. It is at this point that the female frog draws upon what is left of her lipid reserves to make yolks for her eggs. (This is almost precisely the same progression that female eels go through in their reproductive process—a process that can be lethally disrupted by PCBs, as Arjan Palstra discovered.)

The yolks are essentially envelopes of energy derived almost solely from fat, and they must supply her eggs with the fuel necessary for development and growth. The problem here is that when the female manufactures yolks for her eggs, she can also be off-loading dioxinlike coplanar PCB, a ubiquitous and toxic variety of the "magic fluid" that is known to cause embryo deformities that can doom offspring.

Unfortunately, there have been no comprehensive tests on how PCBs might impact frog reproduction specifically since frogs have no constituency like commercial or sport fisherman who depend on them. But since varieties of PCBs are believed to have wiped out entire species of fish such as lake trout and European eels, often in mere parts per trillion, and since vertebrate endocrine systems are essentially alike, there is good reason to believe that the impact of PCBs on at least some species of frogs is equally disastrous.

Of course all this begs the question "Why hasn't this research been done?" The direct answer comes from leading amphibian toxicological researcher, Dr. Michael Fournier of the University of Quebec. He says simply that the "problem with studying the effects of environmental contaminants on amphibians is that there are a limited number of molecular tools that have been specifically developed to study the effects."

Thus the real question as to why the basic research on PCBs and amphibian extinctions has not been undertaken is not a scientific question, but once more a political one. There are no "molecular tools" available to pinpoint the cause of the global amphibian disaster because apparently no nations have given enough money to their scientists to develop them. And so far as can be found, not a single nation, rich or poor, has required that a multinational chemical corporation kick in a dime for amphibian research.

CHAPTER II

The Devil's Gamble

What a strange creature is man that he fouls his own nest.

PRESIDENT RICHARD M. NIXON

The weather for the first Earth Day, April 22, 1970, was appropriate in much of the country, sunny and clear with light breezes—sweet early spring in the United States. Mother Nature seemed to be going out of her way to cooperate with her newly conscious acolytes, millions of whom were taking to the streets of cities across the country, such as New York. There the fashionable mayor John V. Lindsay decided on a classy celebration. He closed vehicular traffic for almost the entire length of Fifth Avenue, where in excess of a hundred thousand celebrants strolled, sang, and were sung to by folk singers and rock groups as Frisbees flew over the heads of academics and their students sitting cross-legged on the sidewalks for "teach-ins" on the basics of the new ecological awareness.

In Washington, D.C., it was much the same, but, as one might expect from place and era, with a harder political edge. Political activists were out by the tens of thousands. And so was the FBI. In the terse language of a G-man, an agent noted in his after-action report that a group of George Washington University students chanted slogans like "Save Our Earth" and one carried a sign with the motto "God is not dead: He is polluted on Earth."

The FBI intelligence information also recorded that Senator Ed Muskie of Maine—an ardent champion of the environment and future Democratic presidential hopeful—gave a short, enthusiastic speech on "anti-pollution" at about 8 p.m. But brief as the speech by Senator Muskie may have been, it set in motion the creation of the greatest set of governmental administrative regulatory actions in history aimed at protecting the planet.

According to members of John Ehrlichman's family, the powerful domestic advisor to the president met with Richard Nixon in the Oval Office the morning after the Muskie speech. Ehrlichman was to later be defrocked and sent to prison after the Watergate investigation, but the morning after the first Earth Day, he shared top lieutenant status with H. R. Haldeman at the Nixon White House. Ehrlichman told the president, "There were twenty million people on the streets for Earth Day yesterday." (It had been the largest public demonstration in the history of the United States and still holds that distinction.) Ehrlichman continued, "Twenty million people and your likely opposition two years from now is Ed Muskie." In his aggressive manner, he kept on hammering the president. "Twenty million people on the streets. Ed Muskie is cued up to take advantage of it and we don't have a piece on the chess board."

"You know," Ehrlichman went on, "there's this idea in it. Pretty easy, sounds like. You take a bunch of water pollution stuff that's currently housed over in the Department of Interior; some air pollution stuff that's in Health, Education, and Welfare; some radiation stuff over in the Atomic Energy Commission. Lump them all together and call it the Environmental Protection Agency. Won't cost a cent because we're already doing all this stuff. You just lump it all together and suddenly you're a player."

So, as the story goes, on the morning of April 23, 1970, the Environmental Protection Agency was born, and coordinated government regulation of pollution was to begin in the United States for the first time. How deep the irony was. While Nixon was being vilified by many millions of activists and liberal conservationists, his administration was in the process of fashioning a revolution in governance that would assure his place in environmental history.

Actually, Nixon had been assessing and positioning himself to take advantage of the popular trend toward aggressive eco-consciousness well before the first Earth Day. Months before, as the centerpiece of his 1970 State of the Union Address, the president had told Congress and the nation of this thoughts and plans for the environment. It was to be, without question, the finest State of the Union Address on the subject of any president before or since. In his most sanctimonious baritone the soon-to-be-disgraced president intoned great words:

> The great question of the seventies is, shall we surrender to our surroundings, or shall we make our peace with nature and begin to make reparations for the damage we have done to our air, to our land, and to our water? Restoring nature to its natural state is a cause beyond party and beyond factions. It has become a common cause of all the people of this country. It is a cause of particular concern to young Americans, because they more than we will reap the grim consequences of our failure to act on programs which are needed now if we are to prevent disaster later.

Sounding far more like Ralph Waldo Emerson than the business-oriented practitioner of realpolitik that he was, Nixon continued,

> Clean air, clean water, open spaces—these should once again be the birthright of every American. If we act now, they can be. We still think of air as free. But clean air is not free, and neither is clean water. The price tag on pollution control is high. Through our years of past carelessness we incurred a debt to nature, and now that debt is being called. We can no longer afford to consider air and water common property, free to be abused by anyone without regard to the consequences. Instead, we should begin now to treat them as scarce resources, which we are no more free to contaminate than we are free to throw garbage into our neighbor's yard.

And Richard Nixon promised that this part of his address to the citizenry was going to be more than just oratory. His administration was going to do something. He would propose to Congress the most expensive water pollution control program in history—$10 billion for municipal waste treatment plants across the country.

Unfortunately, Nixon quickly reneged on budgeting his massive pollution control program. Citing high government spending, the funding for the ambitious program was quietly deep-sixed. It was not really surprising either, for the fact was that even though Richard Nixon's words (perhaps ghostwritten by Ehrlichman) were a stirring call to arms for the environment, the president wasn't personally known to care about ecology in the least.

True to his character, Nixon's environmental philosophy was formed by political expediency. Pollsters' surveys had been recording broad and deep public support for environmental protection in the late sixties. By 1970, the environment was right behind the war in Vietnam in the minds of the public as the country's biggest problem. And of course Nixon, being the preeminent political weathervane of his generation of Washington leaders, grabbed on quick and strong, hoping that his environmental stance would help offset the growing disenchantment with his administration's war policies.

So history left it to John Ehrlichman—that day after the first Earth Day—to come up with a brilliant, simple plan that Nixon could glom on to, which would make him a champion of the environment and, by implication, a man of peace, despite the war raging in Southeast Asia. But not only did Ehrlichman foster the creation of the Environmental Protection Agency, the first governmental agency anywhere solely dedicated to protecting the environment, he also designed its structure and, excepting perhaps the top political appointments, he helped make sure that the EPA was staffed with some of the best and brightest—as many Wayland Swain types as could be found.

Although it is a historical secret, almost all of the Nixon administration's truly remarkable environmental accomplishments were to be the result of John Ehrlichman's bravura bureaucratic performance and commitment, exemplified by his creation of the EPA. How skillful his pro-environment arguments to Nixon must have been, cloaking his advocacy in just the right purely selfish political terms that the president understood, while at the same time avoiding making the paranoiac Nixon fearful of losing the allegiance of corporate America and their cash contributions.

Of course the ending of it all was tragic. While John Ehrlichman must have had perfect pitch to catch the president's ear in ways that would produce, at the time, an administrative symphony of environmental protection, he was tone deaf to the implications of Watergate. Disgracing himself repeatedly at congressional hearings with shameless and bizarre Machiavellian mutterings and demonic smiles, Ehrlichman capped his governmental career as a closet conservationist with a term in federal prison.

Regardless of how eco-historians view the Nixon administration—its accomplishments, motivations, blemishes, and warts—the creation of the EPA was an undeniable milestone in human governance. Be that as it may, it was the United States Congress that passed the first comprehensive environmental legislation the world had seen. The pollution laws that activist legislators of the early seventies put on the books were not only tough, they covered almost every known aspect of pollution control. The Clean Air Act, the Federal Water Pollution Control Act, the Safe Drinking Water Act, the Endangered Species Act, and the basis for the Superfund were all created in the thirty months following Ehrlichman and Nixon's formation of the Environmental Protection Agency.

Out of the orgy of environmental legislation was to eventually come a law inspired by PCBs, a law that would purportedly end the threat of toxic pollution in the United States, the Toxic Substances Control Act. With congressional passage and Nixon's approval, millions of dollars were given to the EPA for basic research on the industrial chemicals that scientists like Wayland Swain were saying had contaminated the nation.

After the banning of DDT for agricultural uses inside the United States in 1972, PCBs became the most high-profile toxic bad boy. When the Toxic Substances Control Act was drawn up and put into force in 1976, it set up regulations for some of the tens of thousands of chemicals that industry had loosed on the United States and the rest of the planet. But the act also contained an unusual series of clauses that dealt specifically with PCBs.

"The new law gives EPA specific authority, which it earlier lacked, to move against an old chemical hazard. . . . One of the most frustrating and long-standing chemical problems we have faced—the problem of PCBs," wrote the then EPA chief Russell Train. Train said the EPA would deal with the PCB problem directly: not only would it prescribe methods of disposing of PCBs, but PCBs would forever be banned from being manufactured under the new regulations.

Russell Train's declaration must have rung a glory chord in the minds of the conservationists of that heady time. But the words of the administrator and the language of the Act would be reduced to nothing more than the facile vaporings of a governmental agency that, although hardly even a bureaucratic toddler yet, was already almost fully co-opted. Enforcement of the tough provisions of the Toxic Substances Control Act was to prove a charade, if not a fraud, from the outset—for hidden in the strong abatement standards of the act was a clause that emasculated it. The clause allowed Monsanto to keep on quietly selling PCBs—even though the Act was trumpeted as a stake in the heart of the contaminant by the EPA administrator himself.

The cunning sham was achieved by including a brief passage that provided for PCBs to be sold as long as they were used in "closed systems." But the fact was that Monsanto had *voluntarily* stopped selling PCBs for uses in "non-closed" systems such as paints, varnishes, and copy paper in the early 1970s, when their own bio-surveys showed that PCBs had contaminated the tissues of just about every living thing in the continental United States that they tested. More importantly, sales to the biggest buyers of PCBs were grandfathered out of regulation by the "closed system" fable.

Since the days of Theodore Swann, the largest electrical companies, such as General Electric and Westinghouse, had always been Monsanto's best customers, using PCBs as cooling and heat-transfer fluids in their transformers and capacitors. It was for those, the most widely used electrical components in the world, that Monsanto had sold millions of pounds—thousands of tons—of PCBs, and now all were specifically exempted from banishment under the Toxic Substances Control Act simply because they were used in "closed systems."

And truth be known, there was no such thing as a "closed system" for PCBs. Capacitors could easily be crushed during landfill operations, leaking their PCBs into the soil or allowing them to be volatized into the air. And transformers, large or small, eventually, even if it took a thousand years, would leak—if, that is, they didn't explode first during weather events such as electrical storms or hurricanes.

Monsanto and top administrators at the EPA surely must have known that the "closed system" rationale was bogus and even potentially deadly. In both the Yusho disaster in Japan and the Yu-Cheng poisoning in Taiwan, the lethal and sublethal doses of PCBs were derived from rice oil manufacturing processes that could have been sold in the United States as a "closed system" under the wording of the Toxic Substances Control Act. And besides that, the Act didn't prohibit the sale of PCBs of *any* sort—"closed" or "open" systems—as exports. Thus, it was legal to poison the rest of the world, if U.S. corporations might want to do so.

Did the head of the EPA, Russell Train, understand the extent of the deception in which he was participating? It is hard to imagine that the importance of the "closed system" wording would have escaped the agency head. Besides, in a curious move for an EPA administrator, Train personally singled out the necessity for continuing the use of PCBs to the national media, saying that if the EPA were to prohibit their use completely, there would be "massive power disruptions." It was, perhaps not so coincidently, precisely the same canard that the Monsanto public relations spokesmen were using at the time.

CHAPTER 12

The Politiks of PCBs

In January of 1972, with 20/20 foresight, Monsanto got its largest PCB buyer, General Electric, to sign an agreement that pledged to hold Monsanto harmless for any liabilities arising from GE's use of PCBs—forever. The exact wording of the agreement was

> Buyer shall defend, indemnify, and hold harmless Monsanto, its present, past and future directors, officers, employees and agents, from and against any and all liabilities, claims, damages, penalties, actions, suits, losses, costs and expenses arising out of or in connection with the receipt, purchase, possession, handling, use or sale or disposition of PCBs by . . . the Buyer . . . including without implied limitation, any contamination of or adverse effect on humans, marine and wildlife, food, animal feeds or the environment by reason of such PCBs.

Why counsel for General Electric would sign such an agreement may seem puzzling at first. Since the late 1930s, GE executives had known that PCBs were toxic and potentially lethal for workers. As well, by 1972, Monsanto was admitting that its PCBs were being found everywhere scientists looked. The implications of such admissions by Monsanto un-questionably would have made PCBs look like a very ugly legal tar baby for GE. Most likely though, it was a confluence of issues that compelled the attorneys for GE to sign the "Hold Harmless Agreement."

Perhaps the most persuasive argument for the GE counsel was that the company's own engineers and scientists were evidently convinced that without PCBs, GE's electrical products, transformers, and the like would become unsafe. This position was not new by any means. Recall that it was taken by GE executive F. R. Kaimer at the PCB summit conference in New York in 1937. Nothing had apparently changed in three decades.

But there were substitute products for PCBs that Monsanto knew about and had, in fact, actually developed. Mineral oil or "high flash point" silicones were available substitutes for PCBs, and some power companies had already started using "dry type" or gas-filled trans-formers—evidently without significant safety problems or high costs. But Monsanto chose not to promote its alternate products, even with its largest PCB customer, because—according to internal memos—it might hurt sales of the extremely profitable line of PCB products.

Another reason that Monsanto didn't push substitutes for PCBs was a calculated marketing consideration. Executives thought that if they tried marketing their newly developed substitutes for PCBs, then their big customers would think it was a sales gimmick. Along with that, there were large European and Asian PCB manufacturers (some with Monsanto licenses) who would be more than happy to sell their PCBs to the addicted U.S. corporations, or at least that was the viewpoint of the Monsanto executives working the problem.

Factoring into their marketing tactics was also their knowledge of the buying behavior of their clients. When Monsanto announced that because of environmental considerations, it would no longer sell PCBs to any customers who used them in so-called "open systems" such as carbonless copy paper and paints, sales of PCBs went through the roof. Stockpiling PCBs that could last them for decades, "open systems" cus-tomers went on such a buying spree that Monsanto could barely keep up with demand and awarded bonuses to executives responsible for main-taining their PCB inventories during the panic run on the product.

To further distance itself from responsibility for its products, Monsanto informed all its clients that since they had purchased the PCBs knowing of the toxic problems the product presented environmentally, they—the customers—would have to figure out a way to dispose of all the PCBs they'd purchased from Monsanto. It was their problem from now on.

When Richard Nixon and John Ehrlichman created the EPA to enforce the slew of environmental laws passed by Congress in the early 1970s, their choice for the first administrator was a cautious, staunch corporate Republican. William Ruckelshaus was one of that group of high-end insiders—princely utility infielders with Ivy League pedigrees—that the executive branch hires to ride herd on the various bureaucracies. But this particular assignment wasn't to be the usual cushy one that a member of that anointed group might usually expect to get.

During his three-year tenure, not only was Ruckelshaus faced with setting up the world's first major environmental enforcement agency, he also had to implement some of the most complex laws ever devised—with tough legislative deadlines. Possibly because of the load placed on him, as well as his ties to business, Ruckelshaus would establish a tradition of executive-level inertia at the Environmental Protection Agency—a molasses-in-January approach toward industry—that would permeate the agency, frustrating if not infuriating some of the EPA's best and brightest, from the inception of the agency until today, almost four decades later. Whatever the reasons for Ruckelshaus's lethargy during his tenure as head of the EPA, he would soon be employed by those he was supposed to police. Approximately three years after his stint as the EPA chief, he would move to the Northwest to become a senior vice president of the Weyerhaeuser Corporation—a company with one of the most rapacious environmental reputations in the country. Later Ruckelshaus would become a proud and longtime member of Monsanto's board of directors, telling the media whenever possible about that company's laudable environmental consciousness, while generally avoiding the subject of PCBs.

Ruckelshaus's successor at the Environmental Protection Agency was Russell Train, another corporate Republican type. While not being literally on the Monsanto payroll (at least publicly) as was his predecessor, Train was apparently enlisted by Monsanto in the public relations battle to save its high-profit PCB product line. And Train was to also continue the generally flaccid approach to regulating industry that had been initiated by Ruckelshaus. One could surmise the result was that Train, like his predecessor, received a hearty reward from the corporate polluters that he was supposed to regulate. After departing from the EPA, Train would become a board member of Union Carbide, the company responsible for the most deadly industrial chemical disaster in world history, India's Bhopal tragedy.

But with President Jimmy Carter winning the presidency on a populist, anti-Washington platform, governmental support for a go-slow approach by corporations toward environmental protection had to go into stealth mode. The new president was not an opponent of corporate hegemony as much as he was a proponent of the underdog. Carter saw pollution and environmental degradation as evidence of immoral business excess, not as an ecological calamity.

Trained as an engineer at the U.S. Naval Academy, Carter didn't have the attitude of a true conservationist like a Teddy Roosevelt or an Al Gore. He also had—perhaps without really perceiving it himself—a Christian fundamentalist view of the necessity of man's dominion over nature, a view that would come to the forefront decades after his presidency with his unwavering support of Monsanto's genetically modified crop programs.

Regardless, Carter appointed Douglas Costle, a sharp Connecticut lawyer with a background in state environmental legislation, as a refreshingly aggressive new head of the Environmental Protection Agency. But as insistent as Costle could be in pursuing tough enforcement policies, his energies were quickly and effectively countered by Carter's Georgia-insider cadre of corporate supporters, including Hamilton Jordan and Stu Eizenstat. In his superior book on the corrosive power of corporate influence on American government, *Who Will Tell the People?*, William Greider quotes EPA chief Costle as saying, "I would say that probably three out of every four [White House] comments on our rule-making were cribbed right from industry briefs."

An Environmental Protection Agency timeline of Costle's tenure would seem to confirm his difficulties—at least as far as PCBs were concerned. Perhaps most telling, in 1979, EPA publicists listed the high points of the agency's accomplishments since its birth in the Nixon years. One of the major EPA triumphs cited was banning the manufacture of PCBs. The trouble with this particular self-congratulation was that it wasn't true. The fact was that EPA chief Costle still couldn't (or wouldn't) remove the "closed systems" clause Monsanto and GE had apparently grafted onto the regulatory package that allowed the continued commercial use of PCBs by the ton.

But while things could have been better under the Carter administration, environmentally speaking, they were about to get worse—much worse—with his defeat for a second term in office.

During the Reagan presidency, EPA pandering to Monsanto would literally reach criminal proportions with the jailing of Rita Lavelle, a special assistant to the EPA administrator, for six months for perjury and obstruction of justice surrounding investigations of regional industrial chemical contamination. One of the primary areas of focus of the Congressional investigation was Times Beach, Missouri, a rural area near Monsanto's PCB manufacturing plant in East St. Louis. According to author Brian Tokar in a 1998 magazine article for the *Ecologist*, citizens' groups uncovered laboratory reports of high concentrations of PCBs apparently manufactured by Monsanto contaminating the soil around Times Beach.

(In an interesting aside, the entire issue of the *Ecologist* was dedicated to articles, mostly done by respected academics, on Monsanto's long-standing global environmental problems. But the issue never reached newsstands in Britain, where it went to press. According to reports, the printer was apprehensive about being sued by Monsanto. British libel laws at the time allowed suits against not only the author and publisher, but the *printer* of libelous materials. Monsanto denied attempting to quash the magazine issue. *The Ecologist*, still in publication, has never incurred a similar problem with their printer.)

In any case, Times Beach area soils were also contaminated with high levels of dioxins—a byproduct of the manufacturing of PCBs. Although the precise cause was never determined, at least forty-eight horses and numerous household pets, along with uncounted hundreds of birds and other wildlife, died, apparently from their exposure to PCBs or their dioxin byproducts. There was also evidence of severe immune and neurological problems in children born to mothers exposed to the contaminated soils around Times Beach. But Monsanto was never tagged as the source of the contamination, even though eyewitnesses said that trucks spraying PCB-contaminated oil on roads were seen picking up oil wastes from the Monsanto PCBs plant.

Not only did the EPA ignore any participation by Monsanto in the Times Beach PCB disaster, the new administrator—Anne Gorsuch Burford—a political operative from Colorado with no environmental background, was ordered by the Reagan White House to withhold EPA documents relating to Times Beach and PCBs from congressional investigators based on "executive privilege."

In a first of its kind, Gorsuch Burford was forced to resign in disgrace as EPA administrator when an investigating panel found that there were serious irregularities in her enforcement of the Superfund. Another scandal erupted when an investigative reporter identified Monsanto as one of the corporations whose executives met secretly with Gorsuch Burford and her special assistant in charge of toxic waste cleanup, Rita Lavelle, who also happened to an ex-publicist for a California chemical company. Lavelle spent approximately six months in prison for obstruction of justice after she was accused of shredding key documents relating to the possible cover-up of Monsanto's responsibility for chemical contamination in Missouri and Arkansas.

Brought into political existence by shrewd, ultra-conservative California businessmen who had groomed him, Ronald Reagan used the disarray and low morale of the EPA as an excuse to gut the agency's funding. By 1983, the White House had chopped away nearly a quarter of the EPA's total budget and gotten rid of one out of five employees, even though the regulatory workload was increasing exponentially.

Not only was the EPA hamstrung by the Reaganites, according to author William Greider, the Reagan White House established a semi-covert executive branch operation where the big-business "fix" could be initiated and carried out efficiently and quietly by going to then vice president George H. W. Bush.

> In Ronald Reagan's White House, it was the office of vice-president that was designated as the chief fixer for aggrieved business interests. Industries that were unhappy with any federal regulations, existing or prospective, were instructed to alert George Bush and his lieutenants. The power of the White House would be employed to intimidate and squelch any regulatory agencies that seemed upsetting to American business.

Greider later quoted C. Boyden Gray, the vice president's counsel, as telling the U.S. Chamber of Commerce, "If you go to an agency first, don't be too pessimistic if they can't solve the problem there. If they don't, that's what the task force is for." Thus under President Reagan, the office of the vice president and the Office of Management and Budget became de facto courts of appeal when Republican big businessmen needed a quick and quiet remedy for troublesome environmental or health regulations.

As far as the industrial chemical contamination of the environment was concerned, perhaps the best example of the efficiency of the Reagan-Bush operation that Greider describes was the destruction of one the most ambitious and important pieces of legislation to come out of 1970s eco-consciousness. Rules for the pretreatment of industrial chemicals, something mandated by the Water Pollution Control Act of 1972, were shelved. The task force chaired by the vice president used one of the most effective of bureaucratic weapons for the killing the provision: they told the EPA to go back and study the problem until they could come up with a better approach—maybe something more comfortable for the chemical industry lobbyists.

When George H. W. Bush became president, he had his vice president, Dan Quayle, take over his old position as the go-to guy for big business. According to Greider, Quayle, working below media radar, did his job quite effectively despite his airhead public image. The vice president was to engineer the killing or elimination of regulations that would have strengthened OSHA, promoted recycling, protected wetlands, and reduced greenhouse gases from power plants as well as their emissions of contaminants like PCBs.

But perhaps Bush's greatest indirect contribution to American eco-political history was the appointment of Clarence Thomas to the Supreme Court. Thomas began his Washington political career as a Monsanto attorney specializing in easing regulatory approval of pesticides through Congressional oversight and enforcement agencies. His lifelong mentor, Senator John Danforth of Missouri (sometimes called "the Senator from Monsanto"), plucked him up after law school at Yale and found Thomas to be a most able representative of corporate interests. Surviving blistering Senate hearings into his morals and ethical character, Thomas was appointed to the Supreme Court of the United States, giving Monsanto a willing ear at the highest level of the third pillar of national government.

As things evolved, Thomas would represent much more than a willing judicial ear for Monsanto. Thomas's wife was reputed to be working for the presidential election campaign of George W. Bush vetting potential appointments when the landmark Gore–Bush election case came before the Supreme Court. Although the issue was hardly covered in the media during those turbulent days, Thomas refused to recuse himself and voted against Gore's appeal. This gave the election to the son of the man who appointed Thomas to the Supreme Court. And it would be the son who, as president of the United States, would later reward Monsanto with a breadth and depth of governmental felicity that some would say made a mockery of the most basic democratic ideals.

Bringing us back to where it all began, in Anniston, Alabama, by the year 2003, Monsanto had divested itself of its Anniston operations (including the long-defunct PCB manufacturing plant originally started by Theodore Swann), which were sold to a company called Solutia. Solutia was the sole creation of Monsanto, and it was easy to guess the reason for the spin-off. Monsanto didn't want its name involved in the $600-

million settlement that the townspeople of Anniston were to receive for living in the most PCB-polluted community in America, if not the world.

Of the twenty-four thousand or so residents listed as the official population of Anniston, twenty-one thousand signed up for suits against the newly formed Solutia for contaminating their homes and their bodies with PCBs. The plaintiffs' lawyers were a spectrum of personal injury attorneys from California, New York, Texas, and Alabama. (The most famous was Johnnie Cochran.)

For their successful efforts, the plaintiff/PCB lawyers would eventually split up amongst themselves some $234 million of the approximately $600 million settlement. Individual plaintiffs—the poisoned citizens—would get cash settlements between $9,000 on the low end and about $50,000 on the upper end of what was left over from the attorneys' take, even though many of them where living with some of the highest levels of PCBs in their blood ever recorded.

But the intrusion of politics in the tragic story of Anniston and PCBs is brought into the sharpest focus when one realizes that this tough Alabama city—suffering from the worst and longest contamination from PCBs of any metropolitan area in the world—has never been designated as a Superfund site by the EPA. One of the lead EPA Superfund attorneys on the matter during the Bush II years, Janet MacGillivray, who later went to work for a Hudson River watchdog group, released a statement through the Environmental Working Group's Web site that gave her perspective on the politics of PCBs.

> The term "Superfund" conjures up an image of a big pot of money, but the reality is the nation's toxic cleanup program is slated to go bankrupt in 2004. [Which it essentially did.—Author] The Bush Administration continues to refuse to fund the Superfund cleanup program, thereby making a conscious decision to safeguard the interests of industry's bottom line over the interests of innocent American families who cannot possibly pay the cost of cleaning up someone else's toxic mess. I've fought GE and battled the need for a clean up of the Hudson River. The people of Anniston have

fought long and hard against Monsanto. The people of Anniston and people who live in contaminated communities throughout the United States have a right to know how polluters have captured this Administration.

CHAPTER 13

The Epiphany

As yet there have been no movies on the life of Dr. Theo Colborn. Perhaps the screenwriters are waiting for history to evaluate the worth of her public warning that PCBs and other synthetic chemicals are having a silent, subtle, but lethal impact on the biosphere through disruption of the hormonal systems of animals—including human beings.

While Dr. Colborn's controversial theories are nothing short of combustible, in the long run, science may well be on her side if the work of her intellectual godmother, Rachel Carson, gives us good measure. Twenty-five years before Colborn started her research program, Rachel Carson, in her consciousness-shaping book *Silent Spring*, told the world that the insecticide DDT was quietly destroying an unknown swath of Earth's wildlife. Even though the chemical industry and its customers fought desperately to maintain their American spraying programs, DDT was banned after the publication of Carson's book—and the environmental movement was born.

An early devotee of the new awareness, with years of local activism under her belt, Dr. Colborn felt she needed academic credentials to truly make a difference. She endured the discriminatory slights that any fifty-year-old graduate student might confront in the world of academia, earning her doctorate from the University of Wisconsin at Madison in 1985.

Soon after, she was hired by a Washington-based environmental "think tank" to study the health of wildlife in the Great Lakes region, which had continued to decline even after supposedly successful clean-ups. More importantly, Colborn was also tasked with finding out what the consequences of the ecological decline of the Great Lakes might be for tens of millions of people who lived in that section of the country.

It was a prime assignment for a new researcher. Wildlife populations had rebounded a bit after the banning of DDT and PCBs along with the imposition of stronger federal water quality standards. Nonethe-less, as the years had passed, scientists from both the United States and Canada continued to document very sick ecosystems throughout the Great Lakes area.

But Dr. Colborn's superiors wanted her to go a step further in her analysis, beyond finding out what was causing the wildlife problems. They believed that the only way to get the attention of the public was to tie wildlife declines to human health issues such as cancer. It was a canny strategy for preserving the environment since one couldn't ring a better alarm bell for the citizenry.

The problem was, the more deeply Colborn looked at the data and studies that were piling up in her tiny cubicle office, the more obvious it became that cancer just wasn't going to be the cause célèbre. Cer-tainly many of the fish found in the Great Lakes around industrial and municipal outfalls were loaded with all sorts of tumors, many of which were cancerous, but the human epidemiological data was contradictory. People living in the Great Lakes region, in both the United States and Canada, though often in highly industrialized settings with heavy air and water pollution, actually had slightly *lower* rates of cancer than the rest of the American population.

At the same time, whole branches of aquatic and terrestrial ecosys-tems associated with the Great Lakes had virtually disappeared since the 1960s. Everywhere scientists looked, species that hadn't quietly be-come virtually extinct were found to have elevated rates of deformities. Field researchers were continually documenting birds with grotesquely crossed beaks and even backwards feet, along with unusually high chick mortality. Even though the Great Lakes supplied drinking water to millions of people without any notable deleterious health effects, biologists found a wide variety of fish species with ballooning livers, misshapen fins, and scrambled or even missing reproductive organs.

Unfortunately for Dr. Colborn, none of the problems was as arresting as those that Rachel Carson wrote about. There just weren't any robins convulsing and dying on the front lawn after a visit by the tree sprayer. No, it was much more understated. Mammals with lowered sex drives, fish with male and female sexual organs, birds with strange-looking beaks: it just wasn't sensational enough for a national news media needing uncomplicated, life-or-death struggles. When Colborn started her work, other than a few handfuls of isolated, nerdy, obsessive scientists (whose papers she pored over) and even fewer focused environmental journalists and writers, nobody seemed to care about the bad things that were happening to the Great Lakes.

In any case, exactly when Dr. Colborn had her insight is unclear, but it must have been a profound personal experience. After she had read and intellectually digested thousands of pages of studies and esoteric field data on the ecological health of the Great Lakes with an impending research report deadline, the light finally turned on. Cancer—terrible as it was—was not the primary issue.

Theo Colborn's scientific epiphany went to the very essence of vertebrate life, the survival of species through sexual reproduction. She had seen, in wide-angled perspective, the secret that Rachel Carson and the best scientific minds in the world had missed: industrial chemical contamination by substances like DDT and PCBs, and their toxic by-products, was either warping or defeating sexual reproduction in large segments of wildlife populations by disrupting or distorting the operation of their endocrine systems—their hormones. And this massive disturbance of vertebrate endocrine systems was the root cause for the continuing devastation of the ecology of the Great Lakes region and—following the same logic—many other parts of the planet's biosphere.

To understand the importance of Theo Colborn's endocrine disrupter conceptualization, you must view cancer not from an anthropomorphic viewpoint, but from a cold, purely scientific one. Certainly the

industrial chemicals—PCBs foremost among them—that polluted the
Great Lakes caused cancer in untold numbers of fish, aquatic birds, and
mammals, as well as contributing to cancer (and neurological disease) in
genetically susceptible human beings. But cancer—in humans and most
other vertebrates—generally develops *after* the organism has had a
chance to breed and pass along its genes. Thus, cancer is unlikely to
cause species extinction because it doesn't usually stop reproduction
completely.

Consequently, it would be improbable that cancer—fearsome scourge
that it may be—could ever wipe out the human race, since there would
be people capable of reproducing *before* onset. In all likelihood, the same
holds true for the rest of the world's vertebrates. As counterintuitive
as it may seem, cancer is, by itself, not a biocidal threat, but endocrine
disruption—at least according to Dr. Colborn's view—is.

Why? To answer that question we need to understand a bit about how
the endocrine system operates in vertebrates. First, like the immune
system that it helps control, the endocrine system is complex far beyond
any complete scientific understanding. Some of the world's brainiest
researchers who have studied endocrinology for decades will admit to
a shoulder-shrugging ignorance of the dynamics of endocrinological
functioning. Some credible scientists even hold that there is no such
thing as the "endocrine system," per se. (They believe that while the
organs of what is called the endocrine system do interact, so do almost
all the other organs in the body in one way or another. Consequently,
they argue that specifying it as a single definable system is misleading.)

In any case, what is known is that the thing which some scientists
call the endocrine system controls and manipulates the vital messag-
ing and signaling functions of cells in all vertebrates from conception
to death. Endocrine glands produce hormones, and hormones are the
body's most important chemical messengers, regulating tasks that must
be performed on a cellular level to maintain life. This includes constant
maintenance of the reproductive, immune, and nervous systems. Hormones are produced in a number of places in the body, including the
pituitary, thyroid, kidneys, liver, pancreas, and testes, to name just a few.
Hormones also control all reproductive processes. It is the hormone

estrogen that the newly forming creature encounters from the mother's glandular secretions, which help to determine how the organism will look, function, and adapt to its environment.

At the same time, most organs associated with the endocrine system are also receptors of hormones. The endocrine system and the hormones that it produces are of an ancient evolutionary lineage. It comes as a shock to most nonscientists, but our human hormones are almost precisely the same hormones chemically that a turtle or a hummingbird uses—and that almost all vertebrates have probably used for many millions of years.

Consequently, the same explanatory metaphors can be applied to all higher animals. When describing the activities of hormones—whether in a human being or an iguana—scientists often use a "lock and key" simile. Though an oversimplification to be sure, the lock-and-key image is reasonably suitable. If the hormone is like the key, then think of the hormone receptor as the lock. When the key finds the lock, then cells within an organ can receive signals to start, continue, or stop an activity.

While hormones are the body's chemical messengers, their message never gets transmitted unless the receptor agrees, and it is here that the theory of endocrine disruption comes into play. It posits that certain industrial chemicals in the environment (PCBs being possibly the single most prevalent and powerful endocrine-disrupting contaminant in the world, although there are many, many others) mimic hormones so perfectly that the receptor is fooled into transmitting a bogus message to the cell. Or the industrial-chemical counterfeit hormone may block access to the receptor by a natural hormone. When false signals or blocked signaling involve a hormone such as estrogen in a pregnant female, then critical instructions to the cells of the forming fetus can be jumbled. In such cases, the impact of endocrine disruption by the synthetic estrogen can range from nothing to catastrophe for the newborn.

But back to Theo Colborn. By the end of 1989, Dr. Colborn had already experienced her "Aha!" moment, in which she realized that endocrine disruption caused by industrial chemical contamination was a grave threat to the long-term biological health of the planet. Now she wanted to bolster her theory and ready it for prime-time scientific and popular acceptance. As part of her plan, Dr. Colborn contacted one of the world's leading researchers in the field of developmental biology, Dr. Fred vom Saal of the University of Missouri. (Recall his criticism of Dr. Stephen Safe in chapter 8.) Dr. vom Saal had been studying the exposure effects of infinitesimal amounts—parts per trillion—of estrogenic substances on mouse fetal development.

With her cross-disciplinary knowledge—along with perhaps a splash of plain old female intuition—Colborn realized that Dr. vom Saal's research provided the possible mechanism by which endocrine disrupters were wiping out a range of animal species in the Great Lakes region. In turn, vom Saal reviewed Colborn's "study of the studies" on the Great Lakes, and what resulted was a series of experiments by vom Saal and his colleagues that caused one of the most heated political/ scientific debates in the last one hundred years.

It had long been a shibboleth of science that "the dose makes the poison." In other words, the toxicity of a substance is generally determined by the amount consumed and the physical size of the subject. (For example, a milligram of cyanide might kill a songbird in a second, but it would be harmless to a blue whale.) In fact, the validity of the dose-makes-the-poison principle, that as you increase the dose of a toxic substance you increase the effect, has been tested, observed, and accepted by scientists for more than two millennia.

But what Dr. Fred vom Saal proved, at least to an influential core of scientists, turned the old rule on its head. Through a series of precise and replicable laboratory experiments vom Saal determined that, under certain conditions with some substances, a low dose could actually have a greater effect than a high dose.

Parts per trillion, incredibly small amounts, of a chemical would produce deformities in the urogenital tracts of the offspring of test animals. But when the same substance—bisphenol A, a global industrial chemical contaminant that is found in plastic products—was given to test animals at doses that were thousands of times greater, comparable deformities were not apparent. Similarly, in vitro tests (tests done in the lab with cell cultures rather than live animals) by associates of vom Saal showed that when human breast cancer cells were exposed to synthetic estrogenic substances at extremely low doses, cancerous cell growth was stimulated. But at higher doses, cancer growth was actually retarded.

What were the mechanisms by which Dr. vom Saal's discovery worked? Nobody knew. But there was informed scientific speculation that it might go like this: When a fetal organism is exposed to a high dose of any toxic agent—a carcinogen or an endocrine disruptor— defense mechanisms are activated through the immune system which then attempt to mitigate cellular damage. But when an organism is exposed to tiny amounts—parts per trillion—of a synthetic endocrine disruptor, the defense mechanisms are not "tripped" because they are designed by evolution to only respond to relatively large amounts of toxins. When dosage is increased to a certain threshold, defense mechanisms finally kick in, detoxifying or attempting to detoxify the chemical threat. At least that's the conjecture in some scientific quarters. (Bear in mind that this process does not prevent the possibility of bogus hormonal signaling occurring concurrently.)

Dr. vom Saal's low-dose theory of endocrine disruption had implications that were nothing short of revolutionary. The core of the upheaval that vom Saal's work represented was that science's central toxicological tenet—"the dose makes the poison"—was wrong. If thousands of industrial chemicals, such as PCBs, were capable of producing (or catalyzing) serious, potentially lethal effects such as birth defects at low parts per billion or parts per trillion, then the entire system of environmental health protection that had been put in place since Rachel Carson's *Silent Spring* had to be totally rethought.

With the popularizing help of Theo Colborn in her book *Our Stolen Future*, Dr. vom Saal's low-dose theory was brought to public consciousness in the mid-1990s and the outcry it created was formidable. The attacks mounted on it, largely by scientists affiliated in some way with chemical manufacturers, were sharp and, in some cases, seemingly reasonable.

The most persuasive argument against low-dose effects causing harm came from Texas A&M's Dr. Stephen Safe. Dr. Safe believed that human beings are directly or indirectly exposed to an enormous variety of naturally occurring hormone-disrupting chemicals, especially those produced by plants called phytoestrogens. These plant-made estrogens are found in myriad foodstuffs such as wheat, oats, rye, rice, soybeans, potatoes, carrots, peas, beans, alfalfa sprouts, apples, cherries, plums, parsley, sage, garlic, coffee, and more. Grains used for making alcoholic drinks such as whiskey and beer are loaded with phytoestrogens.

According to the "Safe argument," human beings are exposed to these naturally occurring endocrine disrupters in ranges from parts per trillion to parts per million in food. Consequently, Dr. Safe, along with the equally renowned and respected Dr. Bruce Ames, the developer of the famed Ames test for carcinogenicity, argued that the massive exposure of humans to natural estrogens at all dose levels hasn't had any measurable impact on human health and in fact may help maintain it.

But as plausible as their attack on Dr. vom Saal's research appeared, it was not without a serious flaw raised by other scientists. The human body, like that of all vertebrates ever studied, retains many of the hormonally active industrial chemicals to which it is exposed. PCBs, pesticide metabolites like DDE, and other synthetic toxic substances can be held in fatty tissue for decades. They also biomagnify up food-chain pyramids, with highly evolved predators having the most contamination.

But naturally occurring hormonally active substances—such as the phytoestrogens found in plants—are not known to be stored in the body for long periods of time locked in fatty tissue in the same way that industrial pollutants are. Nor are naturally occurring substances known to biomagnify to the same extent as industrial chemicals. These facts seem to indicate that most species of animals, probably all, don't metabolize industrial chemicals that mimic hormones in the same way they do naturally occurring ones—even at tiny exposure levels in the low parts per trillion.

Consequently, the arguments of Drs. Safe and Ames, that naturally occurring hormone-like substances and industrial chemical contaminants were equivalent as far as their biological activities were concerned, appear to be weakened beyond good scientific standing by basic observations. Nor did their attack on vom Saal's work acknowledge the fact that though the planet's vertebrates (humans included), over many millions of years, have evolved methods for detoxifying naturally occurring toxins, they haven't had time to evolve defenses against industrial chemicals, given the recent contamination of the planet.

Other detractors, many with industry connections, have attacked Dr. vom Saal's work by saying that his experiments are not replicable. This is an extremely serious charge to make against any scientist, because of the implication of bias and even fraud. The allegations that vom Saal's experiments couldn't be reproduced are particularly rank since, in fact, they have been painstakingly replicated by dozens of other researchers. As of 2005, there had been more than one hundred studies involving aspects of vom Saal's low-dose effects. The overwhelming majority of them (approximately 90 percent) were consistent with vom Saal's findings of significant harmful impacts of endocrine-disrupting chemicals at extremely low doses.

But as the old saying goes, the devil is in the details. It turned out that the source of funding for some of the low-dose experiments seemed to magically affect the results. Of the studies conducted with funding from chemical industry interests, not a single one replicated vom Saal's findings. Eleven out of eleven corporate-oriented research projects found no low-dose effects, or no harm from such effects. The insinuation veiled in the scientific verbiage of some of the industry-linked research was that there really was no such thing as a "low-dose effect," with the pointed suggestion that it was all simply the natterings of chemophobic scientists.

CHAPTER 14

GE and the Jacking of the Hudson

To the handful of people who truly know it, the Hudson River is the most productive, abused, beautiful, ignored, and surprising body of water on the face of the Earth. There is no other river quite like the Hudson, and, for many persons, no other river will do. The Hudson is the river.

ROBERT BOYLE IN *Audubon* MAGAZINE, MARCH 1971

The contamination of the Hudson River by PCBs is America's classic and longest-running environmental drama—a four-decade-long script that has pitted the largest corporation in the world against ragtag groups of conservationists and a Dominican nun. The performance involved the greatest known act of pollution in U.S. history, creating the biggest Superfund site in the nation—some two hundred miles of the Hudson River, arguably the premier estuary on the planet in terms of combined beauty, fecundity, and proximity to humanity.

How John Francis Welch became the CEO of General Electric and the most respected and reviled executive in the country during his reign is a mythic American story. To be sure, Jack Welch was poorly positioned by fate to become the chief of General Electric. From Irish working-class stock, with a Massachusetts mill-town background and a severe stutter in childhood, and rejected by Ivy League schools, Welch just wasn't the timber from which Brahmin GE executives, direct business descendents of Thomas Edison himself, were hewn.

But his close relationship with his devoutly Catholic mother, who made him attend church every day and watched from a front-row pew as he did his altar boy duties, produced a man of will and drive and, as might be expected of such a routine, one of smoldering and sometimes volatile temperament. Welch's mother had brooded, with her inculcation of unbending daily obeisance, an odd duck: a business revolutionary who was a hater of authority, ritual, and many of the normal traditions of corporate life—and the eventual master of them all.

After getting an undergraduate degree at the University of Massachusetts, followed by a doctorate in chemical engineering from the University of Illinois, Welch joined General Electric in 1960 as a junior engineer at the company's Pittsfield plant. There he became involved in the manufacturing and marketing of one of GE's hottest items of the time, plastics. It was a particularly good fit, since the plastics division at GE was probably the only one that would countenance Welch's brash ways and aggressive behavior.

By the early seventies, Jack Welch had done extremely well with General Electric, developing the plastics division into the most efficient and profitable within the corporation. Now he was being groomed as a potential CEO, one of half a dozen candidates. In order to see just how good the plastics wunderkind really was, his superiors decided to assign him the Medusa of company problems: PCBs and the Hudson.

In the fall of 1970, Robert Boyle, a hotshot *Sports Illustrated* editor who was also an ardent fisherman and river rat, had broken a huge story on how the Hudson River was contaminated for hundreds of miles with PCBs. Boyle had gotten the brass at *Sports Illustrated* to pop for a series of expensive lab analyses of fish he personally caught in the Hudson. Boyle's fish had extreme levels of PCBs. But it would take another five years for the source of the massive pollution to be publicly pinpointed, and that would require the help of an aristocratic bureaucrat.

Commissioner Ogden Reid, a scion of the Reid family, publishers of the *New York Herald Tribune*, was an extraordinarily well-connected Republican, a patrician whose roots in the state's political landscape were deep and long. At the urging of Richard Severo, a *New York Times* reporter, Reid announced that scientists working for the state of New York had uncovered the source of the PCB contamination of the great river: General Electric.

General Electric had been using their formulation of Monsanto PCBs at two big plants on the upper Hudson River. The PCBs were being used for a number of applications—everything from cooling fluid in massive electrical transformers the size of sarcophagi to filler in small capacitors. In fact, GE had been blithely dumping PCBs by the ton from their two Hudson River plants since the mid-1930s.

When the *New York Times* broke the story on General Electric's pollution of the Hudson with PCBs in 1975, Ogden Reid vowed that he would force GE to have "zero discharge" into the river as soon as reasonably possible. Having both the publicly declared personal will and the powerful apparatus of government at his disposal, Reid could make it very difficult for GE to wriggle out of responsibility for their apparent ecological misdeeds.

But rather than being scared by the Hudson PCB assignment, Welch relished it. If he could defeat the state of New York, if he could juggle the complex scientific issues, if he could master the arcane legal concepts involved in the convoluted environmental laws that were applicable, if he could triumph as the supreme gladiator in the political arena—he would have the helm of the most powerful corporation in the world reasonably within his grasp. Not only that, the PCB fight would perhaps allow him to bloody and defeat the apotheosis of all he would never be—Ogden Reid, the urbane, Ivy League, Eastern-establishment blueblood.

In any case, it was to be Welch's archenemy Reid who took the first shots of the Hudson PCB war. The indefatigable Bob Boyle had commissioned more surveys of contamination of fish in the Hudson River around the General Electric plants in Hudson Falls and Fort Edward. While almost all the fish tested had much higher levels than the five parts per million safety standard set by the federal government, the PCB tissue burdens of some of the fish were stunning. One of the Hudson River eels was found to have a contamination level of 559 parts per million—an amount high enough that if an adult consumed less than a half-pound portion he or she would be getting a lifetime allowance of PCBs, according to scientists.

With those facts at hand, Ogden Reid did the unthinkable. He closed the Hudson River to commercial fishing for striped bass and eels, as well as other local species. Reid followed up by initiating a state enforcement action against General Electric that would compel the company to totally eliminate their PCB discharges into the river within six months.

Reid's action was not only vehemently opposed by General Electric, but by the New York State Commerce Commissioner, John Dyson, a bureaucrat whose standing in the state government hierarchy was equal to Reid's. Dyson argued that if GE was forced to totally stop PCB discharges in such a short time, it would have to close down its manufacturing facilities, throwing thousands of innocent employees out of work.

With the political heat rising, Reid got a respected Columbia law professor to hold a hearing on the matter of General Electric's poisoning of the Hudson River. Judge Abraham Sofaer listened to hours of testimony, much of it focusing on whether GE had been reckless in its dumping of PCBs into the Hudson. He also got an earful of complaints on the state of New York's decades of stupor in enforcing even the most basic antipollution laws. After evaluating the hearing testimony and evidence, Sofaer issued his report.

The seventy-seven-page opinion that Judge Sofaer wrote was both an indictment of General Electric's corporate behavior and a total vindication of environmental commissioner Ogden Reid's position. "GE has discharged PCBs in quantities that have breached applicable standards of water quality," he wrote. Continuing, the judge opined that GE had also "injured fish, and . . . destroyed the viability of recreational fishing in various parts of the Hudson by rendering its fish dangerous to consume." Sofaer also ruled that General Electric was, as Commissioner Reid had contended, the responsible party for the contamination of the river.

It was crunch time—not only for Jack Welch but for GE. If Welch couldn't beat Ogden Reid and the state, if he couldn't reduce GE's liability for PCB contamination, then the very future of the company was at risk. Literally thousands of civil suits against GE, large and small, could be successful by relying on the validity of Judge Sofaer's ruling. If that happened, then the company would probably be forced to do what the Johns-Manville Corporation did with their asbestos liability and declare bankruptcy. Or at the very least, GE could face a severe stock decline from which it might take many years to recover.

Making matters even more risky, there was no safety net for Welch. There would be no one else to share the blame since General Electric had released Monsanto from responsibility by signing the "hold-harm-less" agreement with them. Nor was there any apparent indication that Monsanto would lend legal or scientific expertise to their old customer for a defense of PCBs. Welch would have to fly solo. But conversely, there was a big upside for Welch: if he were to be successful, there would be no one else with whom to share the glory.

And so the scene was set for Jack Welch, and he would not disappoint in his starring role when the lights came on. As biographer Thomas O'Boyle later wrote, "The state of New York proved no match for Welch."

Taking into consideration the dimensions of the disaster—the contamination of two hundred miles of some of the finest riverine topography inhabited by humankind—Jack Welch, by the summer of 1976, had engineered the most lenient pollution abatement arrangement in the history of New York State, if not the nation.

The agreement between GE and New York State declared that, regardless of the findings by Judge Sofaer, the General Electric Company had done nothing wrong. Further, their legal liabilities would not exceed $3 million. This was a pittance considering that the eventual costs of physically cleaning up the river were estimated to be approximately $1 billion. Even more astounding was that Jack Welch got the state to agree to kick in full matching funds of $3 million in taxpayers' dollars, leaving the clear impression that the state of New York was equally culpable for the PCB catastrophe.

The feisty Irishman also proved the smartest of spinmeisters. After both parties had reached a settlement in principle, Welch forced protracted negotiations on a seemingly trivial point: how GE was to tender their $3 million payment for contaminating the Hudson. The state of New York wanted a check from GE, but Welch would only agree to a wire transfer of the funds. Finally, exhausted negotiators for the state agreed to Welch's demands. When Welch was asked why he forced such a minor matter, he responded that he didn't want government officials to be able to wave a blown-up copy of a GE check in front of the cameras at the news conference announcing the agreement.

It wasn't hard for Ogden Reid to see that Welch was successfully steamrolling the state during the negotiations. Not wanting to be commissioner of the Department of Environmental Conservation when the humiliating results of the accord were announced, Reid fell on his sword and resigned scant days before the Hudson PCB settlement was made public.

While Welch might have taken credit for the political suicide of his archrival, there were other currents that contributed heavily to Reid's defeat. The mid-1970s found New York State in a relatively serious recession. During the PCB negotiations, Welch's superior at General Electric, CEO Reg Jones, personally called the governor of New York, Hugh Carey, and informed him that if the state didn't go along with Welch's demands, then GE would close all their plants and thousands of workers would be put on state welfare rolls—a threat that Carey had heard before but always took seriously.

The governor apparently panicked. There are no records of this, of course, but Carey may well have placated Welch and Jones by giving them Reid's head. Even though Reid served at the governor's behest, Carey was known to be too much of a milquetoast to go directly after his well-respected and politically powerful commissioner. Instead, high-level bureaucratic moles for Carey were mobilized inside the Department of Environmental Conservation with orders to undermine the commissioner at every possible turn. In the meantime, in another message designed to further humiliate Reid, the governor announced to the media that the commissioner's fears were vastly overblown and to prove it, he would drink a glassful of PCBs! But after more cautious consideration and perhaps the recommendation of his physician, the governor reneged on imbibing a "magic fluid" cocktail.

• • •

As Ogden Reid packed his bags, his opponent was being wreathed with corporate laurels by General Electric. Within months, Welch was made a senior vice president of GE. In 1979 he was appointed as vice chairman of the company, and in 1981 Welch reached the position he had predicted he would reach some twenty years earlier, becoming the youngest CEO in the history of the corporation.

With Jack Welch's career reaching its peak, the PCB problem still simmered. Grassroots environmental groups had sprouted to protect the Hudson. Riverkeeper, an organization founded by Robert Boyle, Clearwater, a coalition started in part by legendary folksinger Pete Seeger, and the venerable Scenic Hudson group all began monitoring and publicly denouncing General Electric's contamination of the river with PCBs.

Reinforcing the warnings of the conservationists were numerous scientific studies confirming the toxicity of PCBs for a wide variety of animal species along and in the river. And for the first time, as discussed in chapter 7, research on human PCB toxicity in humans was being made public in the form of investigations into the serious neurological effects of PCBs on the children of women who consumed contaminated sport fish when they were pregnant, as well as the public data from the two lethal mass poisonings by PCBs in Asia. But all the studies and grassroots organizational efforts were vociferously rejected by Jack Welch. He maintained a hard-line tobacco industry response to the PCB problem, repeating a refrain that he would sustain for the next twenty years of his regency at General Electric: that there was no irrefutable proof that PCBs caused any harm to anyone.

In 1980, the Congress of the United States would have its last great ethical spasm of ecological consciousness. Despite intense lobbying by corporate interests, led by General Electric and Monsanto, Congress passed the Superfund law. Though it was complex in form, its essence was simple. The law mandated an eminently reasonable approach to dealing with industrial chemical contamination by substances such as

PCBs: polluters were to be held financially responsible for cleaning up any mess that they created.

Approximately two years after passage of the act, the EPA designated approximately two hundred miles of the Hudson River as a Superfund site. The EPA alleged that GE's two plants had dumped more than a million pounds of PCBs into the river over a thirty-year period. The GE Superfund site on the Hudson was the largest in the United States. (GE was also to be named by the EPA as a responsible party in more than seventy other Superfund cleanup sites around the country, the majority of which involving dumping of PCBs.)

The Superfund designation got Jack Welch motivated. As a regular guest at Ronald Reagan's White House, Welch had the ear of the most powerful man in the land. And to say that Ronald Reagan was a fan of General Electric would be an understatement. Reagan arguably owed the presidency itself to General Electric. When he was an aging B-movie actor whose career was on the wane, GE gave Reagan a TV contract in 1954 to be the host of *General Electric Theater*, a mash-up of pop-entertainment materials. But perhaps not surprisingly, it proved wildly successful, dominating its prime Sunday evening slot. *General Electric Theater* was a consistent top-ten ratings draw for all TV programming, and as GE's spokesman—or the "Face of GE," as the publicists liked to call him—Ronald Reagan honed his political skills, cheerleading for the corporate way to thousands of GE employees all over the country. When *General Electric Theater* ended its run in 1962, Reagan claimed to have visited more than one hundred of GE's plants and research facilities, shaking hands with a quarter of a million people.

Not only was he a more-than-loyal GE booster, but by the time he became president, Reagan had also evolved into a cranky anti-environmentalist who seemed to have developed an irrational hatred for trees. Manning his chainsaw on his California ranch, Reagan would say things like "trees cause more pollution than automobiles" and when you've "seen one tree you've seen 'em all."

Whatever catalyzed such beliefs in Reagan was irrelevant to Jack Welch, but surely the president's anti-environmental fulminations sounded good to the General Electric CEO. The content of Welch's meetings with Reagan at the White House aren't privy to public scrutiny, but one can be sure that Welch gave "his" president an earful on the destructive nature of the Superfund for GE and other corporate polluters like Monsanto.

Reagan listened. Word must have gone out to his EPA administrator Ann Gorsuch Burford that it would be best to enforce the tougher elements of the Superfund program tepidly—if at all. The result, the reader may recall, was—in brief—that Gorsuch Burford got embroiled in a congressional investigation into her administration of the Superfund, withheld inculpating documents, and was cited for contempt of Congress and forced to resign from office—the first, and so far last, EPA chief to undergo such a disgrace. As we've seen, her assistant, Rita Lavelle, who actually oversaw the Superfund, fared even worse.

Next to the Iran-Contra affair, the Superfund scandal, fostered largely by Swann's "magic fluid," would be the worst for the Reagan administration. Nonetheless the result was that General Electric got a pass on having to deal with the Hudson PCB disaster during Reagan's presidency. This was extended to the administration of Reagan's presidential heir, George H. W. Bush. As one would expect from the former go-to guy for corporate interests under Reagan, the White House under George Bush senior produced more of the same governmental lethargy toward taking enforcement actions against GE.

But by the 1990s, hundreds of scientific studies had recognized or were beginning to cite both the carcinogenicity and endocrine-disrupting toxicity of PCBs in minute amounts for wildlife and humans. Making matters even more difficult, the New York State Department of Environmental Conservation—though reduced to semi-eunuch status by its agreement with General Electric a decade and a half earlier—had designated the Hudson River around GE's two plants as a *state* Superfund, matching the federal designation. Worse, the state bureaucrats were recommending a draconian remedy: the dredging of miles of Hudson River shoreline to be underwritten by GE.

Naturally, Jack Welch was ardently opposed to this plan. Even though neutrally funded research studies of PCBs around the GE plants had determined that they represented a serious ecological and health threat to the region that could never be resolved naturally, Welch felt that the dozens of scientists and engineers who backed dredging were just plain wrong. Welch was sure that dredging would just stir up the PCBs in the Hudson and make things worse. As far as he was concerned, all that needed to be done was to let the PCBs remain in sediments—there to get buried deeper and deeper as the eons passed.

Even though it was with PCBs that Jack Welch made his reputation at General Electric, he no longer had time to direct strategy for dealing with the "magic fluid." Rather than handling the PCB situation himself, Welch made a brilliant hire of a man whose competence and knowledge, when applied to manipulating the political and legal systems, would be priceless for Welch.

Stephen Ramsey was a Princeton-educated attorney who liked to characterize himself as just another "boy from Oklahoma." Sporting a salt-and-pepper beard and a mild manner, Ramsey looked like a hip, intellectual tree hugger. Nothing could have been further from the truth. Ramsey was the quintessential Washington Beltway insider, and to some an anti-environmental lawyer with real juice. Not only that, he was more than just a bit familiar with the Superfund; he virtually *wrote* the enforcement regulations for the law himself.

Before being hired by the prestigious mega–law firm of Sidley & Austin, Stephen Ramsey had served as an assistant attorney general in Reagan's Department of Justice at the environmental enforcement section. As a public servant Ramsey toiled to develop and write the rules for enforcement of the Superfund as Congress had mandated. And there were other benefits related to his old government job that made him a highly valuable commodity for corporations like GE with big Superfund problems. As author Thomas O'Boyle pointed out in *At Any Cost*, his biography of Jack Welch, "Besides being a smart lawyer, Ramsey had excellent connections inside the Beltway with the congressmen and administrators who craft and enforce the nation's environmental law."

In any case, it probably didn't take much to convince Welch to hire Ramsey after he learned of his authoring credits on the Superfund regulations. According to freelance investigative journalist Eric Gold-

scheider, soon after leaving his job as an assistant attorney general—
enforcing environmental laws for the people of the United States—
Ramsey "helped write a ten-page manual on how to stymie government efforts to hold companies accountable for pollution." The Ramsey
manual allegedly "included advice on flooding the government with pa-
perwork by using the Freedom of Information Act 'broadly and often'"
and making "artful use of the Book-of-the-Month-Club response," in
other words, "If we don't hear from you, we assume you agree with us."

With the federal government covered, Welch hired another ex–
government official named Peter Lanahan. He was to work the state of
New York, where Governor Mario Cuomo was being less than cooperative with General Electric's efforts to evade responsibility for polluting
the Hudson. Lanahan was, yet again, an ideal insider for Welch. He left
his post as deputy commissioner of the state Department of Environmental Conservation to become a senior manager at General Electric
with an unwritten job description that seemed to consist of two duties:
to baffle the enforcement people with whom he'd just been working and
wage an effective campaign to convince the public that GE was handling the PCB situation with the best interests of the citizenry foremost.

It's well known that one of the easiest ways to stall any regulatory
process in a democracy is to tie up it up with the public hearing process,
a process that is amply provided for in all the environmental laws passed
by Congress. Ramsey and Lanahan were apparently happy to stretch
out the PCB hearings ad nauseam, since every delay meant putting off
having to spend hundreds of millions of dollars on cleanup and dredging. According to O'Boyle, General Electric used "hearings to deluge
the government with commentary, which in turn requires the EPA to
spend months responding to what GE has said." O'Boyle continues:
"Two years after the EPA began the reassessment of the Hudson River
in 1989, the agency issued its Phase One report; GE's response to the
report was more than six hundred pages long, longer than the EPA
document."

Jack Welch's new hires also employed another classic stalling method
used by politicians and corporations: get studies—and then get *more*
studies. According to General Electric's own data from their Internet

Superfund "General Site Information" on the Hudson River, before a reasonable final decision could be arrived at by the EPA regarding PCB contamination, GE engineers and scientists projected that they would need to take thirty thousand separate sediment samples from more than one thousand different locations on the Hudson, and analyze each one for PCBs and possibly other contaminants.

Lanahan also established General Electric's very own quarterly magazine to educate the public on PCBs. It was named *River Watch*, a title evidently intended to confuse GE's propaganda organ with legitimate Hudson River advocacy group publications. But Lanahan went a little too far with this kind of subterfuge. When he inserted the EPA's name as part of a subtitle on the masthead of one of the issues, the agency made an official protest and it was removed. (As part of the deal, EPA attorneys apparently got Lanahan to place GE's name on the front page of the magazine for everyone to see.)

However, being forced to display its true colors didn't stop *River Watch* from relentlessly pushing the General Electric party line on PCBs well beyond what some might consider ethical. Even though it had access to more than sixty years' worth of data on the catastrophic impacts of PCBs, *River Watch* essentially proclaimed that there were "no adverse health effects" from exposure to PCBs.

For Jack Welch and General Electric, as far as PCBs and the Hudson River were concerned, the 1990s presented both public relations and political problems that just had to be toughed out. *New York Times* correspondent Elizabeth Kolbert—one of the most respected environmental journalists in the country—called GE's contamination of the Hudson River "among the most clear-cut and destructive cases of industrial pollution ever recorded." Things were even worse inside the usually friendly Washington Beltway. Welch no longer found compliant bureaucrats minding the store.

In fact, some of the Clintonians were downright unfriendly. In the fall of 1997, Clinton's secretary of the interior, Bruce Babbitt, gave a fiery speech on a scenic overlook on the Hudson River, denouncing General Electric for their handling of the PCB disaster. Babbitt said,

"The fact is, the sickness of this river today is directly traceable to the General Electric Company." He went on to lambaste GE and its management for delaying the cleanup of PCBs and spending millions on lobbyists instead of taking responsibility for the calamity it had created.

A year later, EPA administrator Carol Browner took the extraordinary step of addressing the Committee on Environmental Conservation of the New York State Assembly. "GE tells us this contamination is not a problem," Browner said. "GE would have the people of the Hudson River believe, and I quote, 'living in a PCB-laden area is not dangerous.'"

Browner went on, "But science tells us the opposite is true. In 1996, at the direction of Congress, EPA conducted one of the most comprehensive reviews ever of PCB scientific studies to determine whether the chemicals cause cancer. EPA reviewed more than twenty published, peer-reviewed animal and human studies—conducted by the top scientists in the field. What did the studies conclude? PCBs are a known animal carcinogen and a probable human carcinogen—that the type of PCBs found in Hudson River fish are the most potent of all PCBs."

Ironically, as EPA administrator Browner was publicly questioning Welch's integrity, he was being hailed as a genius and a legend. Just two days before Browner's blistering speech, *BusinessWeek* magazine featured Jack Welch on its cover, billing the story as a "Close-Up Look at America's #1 Manager." The article, which ran fifteen pages, spoke glowingly about GE's profit margins as well as the CEO's "muscular" physique and "laser-blue eyes," while never making a passing mention of the corporation's problems with the Superfund and PCBs.

While the good press accounts always seemed to keep coming, perhaps Jack Welch's worst public performance as chief executive officer was indirectly caused by PCBs, and more particularly by a Dominican nun named Patricia Daly. Sister Pat, as she is known, was a spokesman for ICCR, the Interfaith Center on Corporate Responsibility, which was comprised of 275 mostly faith-based institutional investors who controlled in aggregate nearly $100 billion in U.S. corporate stock.

As an ICCR representative, Sister Pat would go to shareholders' meetings of mega-corporations such as Ford, Exxon, and General Electric and very politely . . . raise hell. Drilling into the company ethos like a dentist eschewing anesthesia for a root canal, Sister Pat would call to public attention some of the more dubious ways and means of CEOs—right in front of the bankers, fund managers, and board members who were backing them. As part of her shtick, Sister Pat would also submit crafty, Trojan-horse shareholder resolutions—a simple process that would invariably drive corporate brass to distraction. (Any shareholder who held more than $2,000 in stock could force the company to include a resolution in their annual report or proxy report that had to be voted on by shareholders.)

In 1998, Sister Pat had submitted a resolution that would require General Electric to launch what she called an "educational program" to warn subsistence fisherman on the Hudson River of the dangers of PCBs. The Dominican nuns operated a variety of schools and outreach programs, including soup kitchens, along the Hudson in places like Jersey City. They knew that many people were eating large amounts of river fish even though the states of New Jersey and New York had been issuing fish consumption warnings regarding PCBs since the mid-seventies, to no avail.

Sister Pat was scheduled to speak about her "educational program" resolution at the General Electric annual shareholders' meeting in Cincinnati. Although there was no doubt that the resolution would never pass, it was the position of the ICCR, whom Sister Pat was representing, that public airing of corporate wrongdoings might eventually produce a disinfecting effect.

With more than a thousand shareholders watching, Jack Welch took center stage and held his audience in thrall with his bottom-line tales and record revenue stats, between rapid-fire quips and jokes à la Johnny Carson. Long gone was the stuttering, rough-as-a-cob, poor Irish kid from a Massachusetts mill town. Welch was jocular, incisive, witty—the quintessential CEO at the top of his game. Except for one minor sour note—a question on executive salaries—all went swimmingly for Welch.

Until, that is, Sister Pat took the microphone. Wearing a trim lady's business suit and coiffed like a successful executive or attorney, Sister Pat told the gathered shareholders that their company had created the largest known PCB spill on the planet and people were being hurt by it. Daly requested that General Electric join with the Dominican sisters in helping to warn and educate the underprivileged meat fishermen who subsisted on Hudson River fish about the real dangers that PCBs presented to them and their families.

Taking back the floor, Welch responded crisply, "PCBs do not pose health risks. Based on the scientific evidence developed since the 1970s, we simply do not believe there are any significant adverse health effects from PCBs." Looking out across his audience, Welch continued, "I want to make it very clear to all of you that we, your company, will base our discussions of PCBs, as we have for twenty years, on science, not on bad politics or shouting voices from a few activists. Advocates can shout loudly. They can say anything. They are accountable to no one."

"Mr. Welch, you are right. We are all accountable," Sister Pat calmly rejoined, "And my accountability is ultimately to God." Then the nun aimed a roundhouse at Welch. "The EPA continues to list PCBs on its suspected-carcinogen list. For you to be saying that PCBs are perfectly harmless is not true. I really want our company to be a credible mover on this." Pausing for emphasis, she continued, "We all remember the images of the CEOs of the tobacco companies swearing that they were telling the truth, swearing that cigarettes weren't addictive. Do they have any credibility in the United States today?"

Reddening, Welch shouted, "That is an outrageous comparison." There was a brief hush in the meeting hall before Daly responded evenly, "That is an absolutely valid comparison, Mr. Welch." Welch replied that "twenty-seven" studies had concluded that there "was no correlation between PCBs and cancer." Again apparently overcome by his temper, Welch yelled, "Sister, you have to stop this conversation. You owe it to God to be on the side of truth here."

Unfazed even by the invocation of the Almighty by the borderline apoplectic CEO, Sister Pat marched rationally onwards. "I am on the side of truth," she said, looking levelly at Welch and ignoring the boos and catcalls of other shareholders. "The other consideration here is that this is not just about carcinogens. We are talking about hormonal disruptions, fertility issues, and developmental problems in children. Those are real issues, and certainly those are the issues that my sisters are seeing in schools all along the Hudson River. That is exactly what is going on here." Realizing he was being outmaneuvered verbally, Welch took cover in his position as chairman of the shareholders' meeting and dismissed Daly with a curt "Thank you very much for coming, Sister."

CHAPTER 15

The Inevitability of Nothing

*To evaluate individual species solely by their known practical value
at the present time is business accounting in the service of barbarism.*

E. O. WILSON, *The Future of Life*

Why George W. Bush would pick Christine Whitman to head up the Environmental Protection Agency was not a mystery. Her Republican pedigree was impeccable. She was a descendent of two of the most powerful political families in New Jersey, the Todds and the Schleys. Her husband, John R. Whitman, an equity investor, was the grandson of New York governor Charles S. Whitman. In fact, Christine Todd Whitman was related by marriage to President George W. Bush through his mother, Dorothy Walker Bush, whose brother's step-daughter married Christine's brother, Webster B. Todd.

During the Nixon administration she worked in the Office of Economic Opportunity under Donald Rumsfeld. Whitman also made productive forays into politics with the Republican National Committee. Soon enough she was a major player in New Jersey politics, nearly beating Bill Bradley for the Senate. In 1993 she became governor of New Jersey, defeating incumbent James Florio by running on a Republican anti-tax platform. Her traditional conservative position, nicely modified for liberal New Jersey, would give her a two-term run as governor, and later the co-chairmanship—with George W. Bush—of the 1996 Republican convention.

Consequently, it must have been a big surprise when Whitman, a compliant and moderate Republican, started pursuing an independent course as the EPA administrator in the Bush II administration. Such a tack naturally ran her right into its most powerful member, Vice President Dick Cheney. Whitman going public with her beliefs on global warming put her seriously at odds with Cheney's perceived position—that to believe the planet was getting warmer was to believe in the effluvium of radicals whose agenda was to destroy the U.S. economy.

Regardless, in a Senate hearing in early 2001, Whitman went forward, telling the senators that "there's no question but that global warming is a real phenomenon, that it is occurring. The science is strong there." Compounding her heretical views were memos for the *Congressional Record* that she had the temerity to send the president, pressing him to address global warming because it was a "credibility issue for the U.S. in the international community."

What followed her testimony, until her resignation, was perhaps the worst series of humiliations of a cabinet member in modern history by a U.S. president. After making a well-publicized campaign promise to reduce carbon dioxide levels when he took office, Bush reneged and went even further, scotching Whitman's announced plans to limit air pollution by big power plants. Then the concept of catastrophic global warming, on which Whitman had lobbied for quick and firm action, suddenly became unacceptable to Bush since Whitman's "science" was suspect. As far as the Kyoto Protocol on greenhouse gases, the president regarded it as more of a threat to the U.S. economy than a pollution panacea, and Whitman was apparently ordered to stop promoting it. As to her promise to conservation groups that the EPA would take enforcement actions against corporations that violated environmental laws, the president was now saying that he thought that there should be a system of "voluntary" compliance by polluters.

Even with the humiliations and the obvious conclusion that her cabinet post and Washington political career were on life support, there was something that Whitman wanted to accomplish, something that she had felt strongly about since she was governor of New Jersey: the pollution of the Hudson River. General Electric was going to have to clean up their PCB mess if it was the last thing she did.

After having been left to fester by Whitman's frustrated EPA predecessor, the General Electric PCB situation was—as always—a political hot potato, but allies in the upper reaches of the Bush White House were nonexistent for Whitman. As well, Jack Welch had unleashed what would turn out to be the most expensive public relations campaign ever mounted by a corporation in an attempt to influence public opinion against enforcement of environmental laws. Nonetheless, Christine Whitman was no political virgin and she had some high cards to play for her last hand at the table. She knew that any administrative decision she made as EPA chief could not be undone by GE publicity no matter how effective their propaganda was, nor could it be reversed by the pro–chemical industry operatives that influenced White House policy. It was Whitman's call to make.

On August 1, 2001, Christine Whitman announced the decision that likely cost her cabinet post. General Electric was ordered to dredge PCBs from the Hudson River—and they were going to have to pay for the removal. If GE refused, then the EPA would undertake the dredging operation itself, with the result that GE could then incur fines that could reach $2 billion—and if GE didn't go along willingly, it would be billed for entire operation anyway.

Conservationists were stunned and elated. Many had believed that Whitman's early expression of concern about PCBs after taking office had been political oratory and little more. Besides, her predecessor, Carol Browner, whose heart was always in the right place and who had the support of the Clinton White House, had effectively left General Electric untouched for eight years. Governor George Pataki, a Republican who had been urging Whitman to take action, said it sounded "very, very positive." New York senators Schumer and Clinton echoed his cautious praise. Even New York State attorney general Eliot Spitzer, the leader who was by far the most knowledgeable on the subject, tipped his hat to Whitman's gumption. (But Spitzer also remarked that while

Whitman was apparently acting in good faith, the EPA was leaving too much wiggle room for a corporation like General Electric with its thirty-year record of bad acting on the Hudson, and the brainy AG's view would turn out to be more than prophetic.)

For Jack Welch and GE, the timing of the announcement by the EPA couldn't have been worse. His retirement party as CEO was scheduled for just a month later, on September 7, 2001. But the forced dredging of the Hudson certainly took some of the luster off the celebration.

Long after Welch's retirement, it would take Sister Pat to establish exactly what the astronomical amounts involved were, but the fact was that Welch had spent hundreds of millions of corporate dollars over the years to defeat any dredging plan. Of course, the basic reason why Welch opposed dredging was that it could eventually cost GE close to $1 billion if they had to pick up the tab, and this would put a serious dent in the company's bottom line. Not to mention that it would be a tough personal defeat at the hands of bureaucrats Welch had dominated so effectively for years.

But for public consumption, Welch opposed dredging not because of the cost to General Electric, but because he was being civic minded. According to Welch's view, dredging the PCBs from the river would stir them up and just make things worse. Certainly, this was a slyly persuasive argument against the EPA plan. Welch and General Electric weren't arguing against the dredging because of costs. Welch and GE were arguing against dredging because it would hurt the environment. According to Welch and his minions, it was those reckless conservation groups who supported dredging that were willing to damage, if not destroy, the river for years to get their way. Welch claimed that it was he and GE who were the real stewards of the river and who put the environment first.

In reality, PCBs in the Hudson had been studied to death for nearly thirty years following Robert Boyle's exposé in *Sports Illustrated*. Dozens of reports by scientists concluded that PCBs were not only causing far-reaching damage to Hudson River Valley ecosystems and the Hudson estuary, but they posed a grave long-term danger to human health in dozens of communities in New York State and New Jersey. Further, and also contrary to Welch's attacks on the cleanup plan, dredging toxic industrial chemicals like PCBs was a relatively precise operation in which stirring them up could be held to a minimum if the appropriate techniques were carefully deployed.

Welch's claim that the Hudson was efficiently cleaning itself and that the PCBs would all soon be locked in the sediments of the river, was contradicted by every study done by neutral academic and state-funded researchers. The fact was that PCBs weren't permanently sequestered in the sediments of the Hudson, but were constantly leaching back into the river in large enough quantities to kill wildlife and contaminate fish and human beings for hundreds of miles and hundreds of years.

Less than a week after Jeff Immelt took over from Jack Welch as CEO of General Electric, in early September 2001, the World Trade Center was attacked and GE stock plummeted, forcing Immelt to react to events rather than orchestrate them. A year younger than Welch had been when named to head the corporation, Immelt had to wait more than three years after 9/11 to find his sea legs and demonstrate his own particular executive *oeuvre*. After a global marketing study, Jeff Immelt announced that General Electric would have a new corporate advertising slogan, "Ecoimagination," and the new company motto would be "Green is green." The cross-media ad campaign featuring pictures of trees growing from smokestacks and comely female miners working in eco-friendly coal mines, would cost GE somewhere in the neighborhood of $90 million. But the steep price was going to be worth it, since the new corporate environmental consciousness was to inform every product that General Electric made, from jet engines to light bulbs.

And the new philosophy of the CEO of the second most valuable corporation on Earth (Exxon had by now taken over the number one spot) wasn't the nattering of a divinely inspired tree hugger. Immelt was always honest about his perspective and motives. He explained that actually, "Ecomagination" was strictly in keeping with the central creed of capitalism, that his corporate redirection was aimed at making profits.

What Immelt wanted to do was take advantage of what he saw as the biggest corporate opportunity to bolster revenues that the twenty-first century would offer General Electric: making products that would save energy and therefore cut costs for everyone from large corporate customers to welfare mothers. Regardless of its radicalism, the business media uniformly praised the new General Electric policy, citing Immelt's vision and courage. Conservationists, though wary, were also pleased with GE's turning over a new "green" leaf.

But Jeff Immelt's "green" vision would become very blurred, if not blind, when it came to the Hudson River and the dozens of other Superfund sites that General Electric had contaminated with tons of PCBs. The problem was that cleaning up PCBs hurt the bottom line. While it might be profitable to make consumer products that were energy efficient, and it might be highly profitable to manufacture big-ticket industrial systems to help other corporations reduce their greenhouse gases, it definitely was *not* profitable for General Electric to clean up the environmental messes they had created around the nation and globally.

So while Immelt styled himself as one of the staunchest allies conservationists had in the corporate world, his position toward cleaning up PCBs would be even more potentially disastrous for the environment than Jack Welch's in-your-face stance. Not only would General Electric get an indeterminate postponement of Christine Whitman's order for cleanup of the Hudson River PCBs, the company would threaten to virtually destroy the Superfund itself on Jeff Immelt's watch.

To be sure, General Electric's legal assault on the Superfund was started by Jack Welch. It was part of his strategy to fight the cleanup of PCBs on a multi-front basis. If by some chance GE couldn't evade the EPA order through political pressure, then it would fall back on judicial remedies of the most potent sort—an assault on the very constitutionality of the Superfund law itself. This would be expensive for GE, but Welch was never one to skimp when it came to fighting enforcement of PCBs and toxic contamination laws. He would hire the finest to deal with the EPA and the troublesome conservationists.

Just before passing the baton to Jeff Immelt, Welch bought the best that money could buy, perhaps the most famed constitutional law expert in the United States, attorney Laurence Tribe. It certainly was a counterintuitive choice on the face of it. Tribe was known for his devotion to protecting the rights of individuals, for fighting—like a modern-day Clarence Darrow—for the underdog, the underprivileged, the oppressed, against the monolithic, entrenched, uncaring establishment.

But motives are frequently difficult to ascertain. Whatever his reasons, he was apparently willing to relinquish his reputation as a progressive-liberal hero and launch an attack for a mega-corporation with a long history of violations of law, and all this against the nation's most important and fragile environmental legislation, the Superfund.

Using the platform provided by appealing a lower federal court ruling that went against General Electric, Tribe mounted a powerful challenge to the Superfund that coincided—and collided—with Jeffrey Immelt's pronouncements in February of 2002 that GE was going to do its duty and clean up the Hudson. Two weeks after pledging to cooperate with the EPA in removing PCBs from the river, Immelt had apparently quietly given GE's legal action against the EPA and the Superfund the go-ahead.

With Tribe's abilities at crafting the strongest of constitutional arguments, General Electric's assault on the Superfund was the most significant and potentially damaging that the law had faced since its passage in 1980. Tribe argued that the EPA's enforcement of the Superfund—ordering GE to "unilaterally" clean up the PCBs in the Hudson—violated the due process clause of the Fifth Amendment.

To some, the case that Tribe so persuasively built for his employer was a sad and shocking perversion. The Fifth Amendment went to the very heart of the Constitution. It was originally intended by the founding fathers to be a central part of the precious bulwark against the government acting capriciously and unreasonably regarding the rights of citizens—people—*human beings.*

But this basic protection against tyranny was twisted by the Supreme Court in the early twentieth century when a majority of the justices, agreeing with attorneys representing western robber barons, held that corporations had the same rights as people. In fact, in the eyes of the Supreme Court, for most constitutional intents, corporations *were* people. And it was on the basis of that bizarre legal aberration, decried by constitutional scholars for decades afterward, that Tribe mounted his attack.

For the media, Tribe justified the General Electric attack on the

Superfund by couching it in populist terms. Ignoring the irony that he was receiving large fees from the second-richest corporation in the world at the time, Tribe exclaimed, "Just as every person in America faced with a government order has the right to appeal to an impartial judge, a company faced with an order from the Environmental Protection Agency to undertake a project of unlimited scope and duration has the right to a timely hearing in front of an impartial judge."

Naturally, Tribe's arguments incensed attorneys for conservationists who were supporting the EPA through amicus (friend of the court) briefs. At the Natural Resources Defense Council, senior attorney Katherine Kennedy felt that the GE suit—if it was successful—might mean for all practical purposes that the PCBs in the Hudson River would never be cleaned up. "[If] the court allows polluters to sue before cleaning up, GE would then use its vast resources to delay the cleanup of the Hudson River for decades more."

The stakes, in actuality, were even higher than that. If General Electric won the case, not only could conservationists forget about seeing PCBs removed from the Hudson River in their lifetime, but the entire Superfund law would be put into jeopardy. Tribe vehemently denied doing this but the appellate judges reviewing Tribe's challenge believed that this was precisely his intent. They wrote, "GE's due process challenge to the [Superfund] administrative orders regime is not a challenge to the way in which EPA is administering the statute in any particular removal or remedial action or order, *but rather it is a challenge to the statute itself* [emphasis added]."

So there it was. If Jeffrey Immelt's newly "green" GE, abetted by Laurence Tribe, prevailed, then not only would it be likely that General Electric could get out from underneath its seventy-plus other Superfund sites around the country (many of which were contaminated with the most toxic and persistent types of PCBs), but all the other corporations with Superfund sites could tramp through the legal hole that Tribe had blown in the law.

On March 2, 2004, the U.S. Court of Appeals for the District of Columbia ruled in favor of Laurence Tribe and General Electric. It was ruled that the federal district court judge had erred in denying GE's right to challenge the Superfund law using the due process clause, and the case was remanded back to the erring judge for reconsideration. With that, the Superfund was legally put into play. While Tribe's apparently potent appeal for GE ground its way through the federal court system, the Superfund law would be fair game across the country for attorneys representing corporate polluters.

On October 6, 2005, after extended negotiations with the EPA—under a new and more understanding administrator following Christine Whitman's resignation—General Electric finally committed to the dredging of PCBs from the Hudson River. The *New York Times*, long an editorial supporter of dredging, enthusiastically reported, "Nearly three decades after PCBs were discovered in the upper Hudson River, General Electric made a binding agreement yesterday to dredge them from the river in one of the largest and most expensive industrial cleanups in history."

But it was too good to be true. Within months, the prescient New York attorney general Eliot Spitzer and his team found easy loopholes and large footholds for more General Electric delays in the binding agreement—the worst of which seemed to be that the EPA didn't actually oblige GE to complete the job. According to Spitzer, the EPA had failed "to require GE to complete removal of PCBs from the upper Hudson River by allowing the company to opt out of the PCB cleanup after the first year, during which time only about ten percent of the PCBs slated for removal would be dredged."

Spitzer noted that giving General Electric such an out was contrary to the requirements of the Superfund law itself, and besides that, the EPA had never before given a polluter that kind of astonishing legal leeway. Equally bad was the fact that by breaking down the cleanup into stages, the EPA had, in effect, prearranged for General Electric to have all sorts of new prospects for delay because of the public hearings and review processes required by the Superfund.

The Natural Resources Defense Council was also withering in its criticism of the agreement between the EPA and GE. NRDC attorneys felt that if the EPA allowed General Electric to commit to only a partial cleanup, then there could only be two outcomes: either the taxpayers would have to foot most of the billion-dollar cleanup bill or the PCBs would remain in the Hudson River.

There was worse to come. Not only had General Electric wrung a seemingly ultra-sweet deal from the EPA, but the agency itself was now acting like a corporate protector rather than enforcer of environmental laws. The *New York Times* published an internal memo from a sister agency, the National Oceanic and Atmospheric Administration (NOAA), in which its knowledgeable pollution experts told their EPA colleagues that their agreement with General Electric was grossly insufficient for dealing with the PCB contamination of the Hudson River. But the EPA had suppressed the memo, keeping it off the public record, along with many other critical memos from state and federal agencies, and even some from the EPA's own staff. (All were brought to light by watchdog groups using the Freedom of Information Act.) Attorneys for conservationists believed that there were more damning documents, possibly even revealing collusion between certain EPA officials and General Electric. But regardless of whether those materials would ever come to light and corrective action be taken within the EPA, the Natural Resources Defense Council lead attorney on the issue, Larry Levine, was not sanguine about the long-term prospects of General Electric undertaking a real cleanup of the Hudson River PCBs.

Levine told reporters, "From where I sit, GE is first and foremost a giant profit-maximizing enterprise. That's a great thing for the development of clean technology, and by extension the climate. But when it comes to cleaning up the messes they made in the past, they lack both profit motive and, so far, environmental commitment."

After ten years of filing shareholder resolutions on behalf of religious investors, Sister Pat Daly finally got something of value out of General Electric. She announced on January 10, 2006, that GE had released documents she had been requesting since the Jack Welch years—documents that revealed the amount of money the corporation had spent fighting the cleanup of PCBs.

According to the figures that Sister Pat posted on the Internet, from 1990 to 2005, at three locations heavily contaminated with PCBs—the

Hudson River, the Housatonic River in Massachusetts, and the Rome, Georgia, area—General Electric had spent a total of $799 million, for the most part stonewalling a full PCBs cleanup. More than $100 million was spent on lawyers and public relations; $40 million was spent on staff salaries and "other overhead." The amounts were generally carried as "unaudited" by General Electric—an accounting euphemism assumed to mean that it could be more, or it could be less.

In any case, the real reasons that CEO Jeffrey Immelt decided to release the PCB expense figures can only be imagined. He may have thought that it would garner some public sympathy for General Electric and the onerous burden they shouldered having to deal with their PCBs. It is doubtful that Jack Welch would ever have released the numbers, if for no other reason than it would have meant open humiliation for him at the hands of the gently fearsome Sister Pat.

But now they were there for the entire world to see—all the appalling costs of trying to duck responsibility for PCBs. If Immelt had hoped he'd gain some public relations advantage for General Electric by making its PCBs expenses public, he would be unhappily mistaken. Sister Pat would ever so politely stuff the figures down GE's throat. She pointed to the undeniable reality that the corporation had spent—at the very least—$800 million and not a single molecule of PCBs had been dredged from the Hudson River. According to Sister Pat, GE had already spent hundreds of millions of dollars more fighting PCBs than the *entire cleanup* would have cost them.

"The bottom line is clear," wrote the Dominican nun. Eight hundred million dollars "is a staggering amount of money and an astonishing use of corporate funds to postpone the inevitable: The fact GE will have to clean up their toxic PCB pollution." It was an optimistic position held by a strong woman of deeply held moral and spiritual convictions—but not one that jibed that well with history. If Jack Welch and General Electric had proven anything over the last half century, it was that when it comes to PCBs and the Hudson, *nothing* was inevitable.

As these words are written the Hudson River is being dredged for PCBs. Hundreds of rail cars full of contaminated Hudson River sediments are being trans-

ported to West Texas as part of the final EPA "disposal" deal with GE. (The dredged material is considered so hazardous that the EPA has refused to release the train route because of the possibility of terrorism.) The contaminated sediment is to reside in the waste facility of an elderly multibillionaire—Harold Simmons, a heavyweight Bush supporter who paid for the controversial "Swift Boat" political ads used against John Kerry and, according to published reports, spent nearly $3 million on a campaign attempting to link Barack Obama to a sixties radical during the 2008 campaign.

Simmons's waste facility, where the PCBs are to be stored in perpetuity, is located above the giant Ogallala Aquifer, arguably the most important agricultural aquifer in the Western Hemisphere. Hydrological surveys indicate that groundwater is already within feet of the PCB holding area. Incredibly, the EPA required no environmental impact statements for the greatest transfer of toxic industrial chemical contamination ever undertaken. Given the behavior of the "magic fluid," if even small quantities of the Hudson River PCBs were to leak into the Ogallala aquifer, experts say it could conceivably be one of the greatest ecological disasters the nation has ever experienced.

CHAPTER 16

Precautionary Agonistes

*The science that supports the Risk Paradigm—the same science that assures us
that low-level contamination does not threaten public health—is intrinsically
linked to that framework's political values, assumptions, and commitments.*

JOE THORNTON, *Pandora's Poison*

A deceptively simple concept called the "precautionary principle" appears to be destined to profoundly change our perspective on human
health and the health of the biosphere, if it takes cross-cultural hold.
Thought of as new by its present disciples, it is anything but. The precautionary principle was first embraced and made doctrine 2,500 years
ago by the father of modern medicine, the ancient Greek physician-phi-
losopher Hippocrates, as part of the physician's oath to "do no harm."
However, what does make the precautionary principle new and radical
is that its adherents now want to apply its rules not only to physicians,
but to global industry.

Originating in West Germany during the late 1970s and early 1980s,
the precautionary principle movement was catalyzed by the recogni-
tion of the gradual destruction of the Black Forest by acid rain coming
from large power-generating facilities to the west of that fabled region.
Private landowners and 'greens' besieged the German government to
reduce plant emissions. Their pleas were heard and a crash effort was
undertaken to save the Black Forest, with a good measure of success.

The legal/philosophical underpinning of the movement—*Vorsorge-prinzip*, roughly translated as the "foresight" or "care" principle—would become known internationally as the precautionary principle (generally called the "precautionary approach" in the United States).

In essence, the precautionary principle was the "do no harm" dictum translated to the practitioners of heavy industry and chemical manufacturing. It also went a step further than the Hippocratic oath in requiring potential polluters who wanted to release any type of contaminant to reasonably establish that their discharges would not harm the environment or threaten public health. In practical terms, this means that rather than being able to use the biosphere for global chemical experimentation, corporations must assume the burden of proof to determine the safety of their products *before* putting them on the market.

Within a few years of the Black Forest success, the precautionary principle had garnered surprising political purchase and had been included in a series of international treaties and protocols. Starting with an agreement to reduce chemical pollution in the North Sea, the precautionary principle gained multiple endorsements as an acceptable precept of international law. In 1987 it was a part of the Ozone Layer Protocol. Three years later, despite difficulties, the precautionary principle was one of the key issues that participating officials agreed on and included in the Ministerial Declaration issued by the Second World Climate Conference.

In 1992, at the famous Rio Earth Summit, the precautionary principle was easily the most important legal/philosophical concept discussed at the conference. It was concisely stated as Principle 15 in the Rio Declaration on Environment and Development. "In order to protect the environment, the precautionary approach shall be widely applied by states according to their capabilities. Where there are threats of serious or irreversible damage, lack of full scientific certainty shall not be used as a reason for postponing cost-effective measures to prevent environmental degradation."

Not surprisingly, given the stature of the Rio Earth Summit, a year and a half later, at the Maastricht Treaty conference that established the European Union, the member nations declared that the precautionary principle would be *the* guide for fashioning EU environmental and health policies. And in 2001, the Stockholm Convention on Persistent Organic Pollutants cited the precautionary principle as a central tenet of policy making for the control of industrial chemicals.

In the United States, it would take a few more years for the concept to catch on with conservation activists, but catch on it did at the seminal Wingspread Conference held in Racine, Wisconsin, in 1998. Convened by Theo Colborn and her colleagues because of her landmark research on how endocrine disrupters were ravaging whole populations of North American wildlife, the Wingspread Conference defined the precautionary principle in complete form for the first time, including its four major points: *acting on early elements of harm, shifting the burden of proof, exercising democracy and transparency, and assessing alternatives.*

With the best definition so far of the precautionary principle on hand, conservationists and public health defenders had something to rally around 2500 years after its first application in principle. Now they had a code, a belief—something they could be for, not against. The precautionary principle gave the international environmental movement, long saddled with accusations of obstructionism and negativity, a positive banner to coalesce under. Best of all, it was a simple, understandable concept with necessity behind it. Just what would necessitate—and catalyze—such a radical, but at the same time rational, approach to ensuring the long-term health of the planet? . . . Nothing less than Ted Swann's "magic fluid."

To use a hackneyed but apt metaphor, PCBs were the perfect chemical "poster child" for the precautionary principle. According to Dr. Peter Montague, director of the Environmental Research Foundation and publisher of the newsletter *Rachel's Democracy & Health News*, our system of protecting public health in the United States couldn't be faulted for "failing" with the PCB catastrophe. It was working "just as it was designed to work." Continuing his 2003 speech at the Fairfield University conference on PCBs, Montague told his audience, "Our government does not seem to have learned anything from the PCB disaster. Our government is now promoting powerful new technologies without regard to future consequences. It was this 'growth at any cost' mentality that brought us the PCB disaster. It is up to us, the citizenry, to insist that our government adopt a new way, a 'precautionary approach' to new technologies."

What led to the systemic failure of governance, documented by Montague and others, that allowed the PCB disaster to occur and, in turn, helped lead to the rise of the precautionary principle? More and more experts are pointing to the breakdown of a scientific method called risk assessment.

Risk assessment is a system, a methodology, for analyzing risk. It came into widespread use in the 1970s when engineers began using it to analyze the safety of structures such as bridges. Risk assessment was an extremely powerful tool, especially when coupled with computer data-crunching capabilities, for producing a solid understanding of margins of safety in the building of physical structures. Successful as it was, by the 1980s risk assessment was being applied to chemicals manufacturing by well-intentioned civil servants looking for methods of making decisions founded on science.

But risk assessment was (and still is) based on the assumption that Earth and the human beings on it have "assimilative capacity." What this means is that risk assessments take for granted that there is a certain level of industrial pollution that can be assimilated without any harm. This assumption—that assimilative capacity can be precisely determined—then leads to a second assumption, which is that once we know the assimilative capacity of Earth for a single specific industrial chemical, we will be able to control damage from it or at least be able to make the risks known.

But if we examine these assumptions more closely, we see that while sounding "scientific," they are actually *political* pronouncements that have been injected into our belief system. The fact is, there is no hard-and-fast scientific definition of assimilative capacity. Assimilative capacity is a subjective designation based on some undefined interpretation of scientific data by individuals. Furthermore, our experience with PCBs tells us that even if we could know the true assimilative capacity of

Earth, we cannot control the vast damage done by the release of rogue industrial chemicals.

The argument above is on the theoretical side, it must be admitted. Perhaps to get a better feel for the realities of risk assessment we should delve into its actual application with a scientist. To do that we'll hear from Oregon-based author and biologist Mary H. O'Brien, in her speech "Beyond Democratization of Risk Assessment": "Twenty-one years ago, I believed that the preparation of risk assessment was a good idea, 'sound science,' as it were. I believed that all that was necessary to correct their [the chemical industry's] frequent downplaying of risk was insertion of more accurate numbers. But I soon became disavowed of this belief."

O'Brien goes on to detail two examples of how assessments she analyzed had gone wrong. In the first, she "became involved in a suit against the federal Animal Plant Health Inspection Service for their nationwide environmental impact statement for aerial insecticide spraying for gypsy moth. The EIS included risk assessment for four insecticides, and for a deposition, I analyzed the risk calculation for the carbamate insecticide carbaryl. There were about ten elements to the risk formula, and I found that each one was an estimate, based on some kind of guess or literature. But as I looked at the scientific literature, I found reason for more conservative numbers for each of the ten elements.

Wrote O'Brien: "For instance, the estimate had been made that five percent of carbaryl would be absorbed through exposed human skin. I found a study using human volunteers (and I do not approve of such studies) by Howard Maibach of UCLA in which he had documented over 70% absorption of carbaryl through human skin. So I re-analyzed the entire risk assessment for carbaryl, showing that with more valid numbers in a number of the model's elements, the risk of harm could be orders of magnitude higher. The judge dismissed this claim of our case, saying I was 'nitpicking.' (We did win the case on other grounds, however.)"

Another case involved an assessment for the herbicide dacthal, which was contaminating drinking water in Oregon. The safe standard for dacthal had been set twenty years earlier based on a rat study using an inadequate number of subjects that had, in any case, been compromised by the fact that they were being given an antibiotic regimen at the same time (for lung infections). The study found no safe levels of the chemical, and tumors in all of the subjects.

Yet the safe standard "had been calculated as if the lowest dose had not caused any adverse effects (which was not true), the tumors had not been observed (which was not true), and a 20-kilogram child were drinking the water," wrote O'Brien.

A year later, the EPA set the standard at even higher levels. "Why did dacthal suddenly look seven times safer?" asked O'Brien. "Had a new study emerged? No, the EPA was still using the same 1963 rat study, but now the formula assumed that only 70-kg adults, not children, would drink the dacthal-contaminated water," *and* it still ignored dacthal's tumorigenic effects in insisting there was no risk of cancer.

"At this point, I realized that life was too short to spend it trying to assist with making better risk assessment when clearly, the goal of most risk assessment, was not to think about harm that could occur, but to instead establish some standard for how much of the hazardous substance or activity would be considered 'safe' or 'acceptable.'"

And there is the nut of it, the reason why the chemical industry and its clients were, and still are, so enamored of risk assessment. Risk assessment was a way to pollute, a method—seemingly scientific—that was actually comprised of arbitrary and subjective judgments that would allow nearly unlimited possibilities for discharging or dispersing their products with minimal regulatory intervention.

As you've just read in Mary O'Brien's account, subjectivity, the lubricant by which decisions can be made to accommodate almost any interest, was infused into the risk assessment process in spades. For the most part, chemical risk assessment typically involves nearly two dozen careful, tedious steps to determine variables like an organism's uptake of a chemical, the varying ways that a chemical is transported, the actual magnitude of discharges, the substance's persistence in the biosphere, its bioaccumulation profile, its toxicity for a variety of lab animals from mice to monkeys, and more.

Then the scientist must decide if these factors, alone or in combination, present a risk to the environment or public health. To do that, he or she must subjectively choose from either ranges of data that have been accrued, or often simply accept arbitrary safety factors created by regulatory agencies, as Mary O'Brien was faced with.

Given all the hurdles and tedium, not to mention monetary costs, it is no wonder that only one hundred or so chemicals, of the almost eighty thousand in the EPA's Chemical Substance Inventory, have undergone anything even near a thorough risk assessment since analysis of toxic materials was mandated by Congress in the 1976 Toxic Substances Control Act.

Thus we have another reason why the risk assessment process suits the chemical industry, since there is a backlog of tens of thousands of chemicals that have not undergone anything near complete testing. To make matters even sorrier, on average, there are seven new chemical applications being filed every day at the EPA, and clearly at those rates the EPA will never catch up with its risk assessment duties. Worse yet, mandated EPA risk assessment under the Toxic Substances Control Act doesn't even include some of the most important categories of chemical contaminants. Eight product categories are exempt from regulation by the EPA under TSCA: pesticides, tobacco, nuclear material, firearms and ammunition, food, food additives, drugs, and cosmetics. Most fall under the purview of other government agencies such as the FDA. (It's also simple for corporations to keep toxic chemical information from public scrutiny by using an easy out provided by the EPA: virtually all they have to do is claim confidentiality when filing a submission.)

While the EPA is shorthanded, and ill funded to carry out its duties in protecting the public and the environment, risk assessment has become a cottage industry in and of itself, with major universities funding risk assessment research for multinational chemical corporations like Monsanto and Dow. As Peter Montague puts it, "It is no exaggeration to say that the modern industrial system with its culture of 'growth at any cost' could not maintain its present course without risk assessors to run interference."

The chemical industry and its customers tout risk assessment as part of a need for "sound science," but when it comes to analyzing their products, in reality, risk assessment, whether done by corporate researchers or by regulators is anything but definitive in its current version. As Montague points out, there are more than a dozen places where risk assessment as it is currently practiced fails miserably in determining the safety of commercial chemicals. In abbreviated form, here are some of the factors that, according to Montague, confound risk assessment:

- Risk assessors try to account for human variability by applying a "safety factor" of ten to their numerical estimate of risk. But such a number has nothing to do with science—it is a guess.
- Risk assessment of chemicals is conducted on single chemicals, but in the real world we are all exposed to mixtures of chemicals day in and day out. Furthermore, many studies have now shown that harmless amounts of individual chemicals in combination can add up to a quite harmful dose.
- Traditionally, chemicals have been tested at the highest doses that laboratory animals could tolerate, but now we know that high-dose tests may miss important toxic effects that occur only at low doses.
- We now know that cells respond differently to chemicals depending on their prior history of exposure. In addition, whole organisms (mice, humans) exhibit similar behavior: response to a chemical is strongly conditioned by prior exposure.
- Thousands of potentially important biochemical reactions are ignored during risk assessment. Even when animal tissues are examined under a microscope, not all tissues types are examined. All organs are composed of various types of cells, and each type would need to be examined to claim that a thorough investigation had been conducted, but this is not done.
- Important routes of exposure are typically ignored—such as inhalation and absorption through the skin.
- With rare exceptions, the period of greatest vulnerability (corresponding to the human period of life from conception through

age eighteen) is not tested in laboratory animals. Adult animals are tested. In addition, effects on second and third generations are not typically looked for.

- Some cause-and-effect relationships between industrial contamination and disease will most likely never be established because causes and outcomes are multiple, latency periods are long, timing of exposure is sometimes critical, and unexposed control populations do not exist.

When so many complicating factors remain unidentified, when so many safety dynamics are not tested, when so many arbitrary, unfounded rules are applied, then risk assessment becomes not a method for determining safety, but little more than an expensive, time-consuming exercise in irrelevant scientific subjectivism. As author Michael Pollan put it with graceful simplicity in his 2001 guest editorial for the *New York Times* on our misguided love affair with risk assessment, "Whatever can't be quantified falls out of the risk analyst's equations, and so in the absence of proven, measurable harms, technologies are simply allowed to go forward."

Pollan's view would certainly apply to industrial chemicals as well as technologies. Just how many of these synthetic substances have been "allowed to go forward" without proper safety vetting for humans, wildlife, or flora? Nobody knows—but estimates are nothing short of staggering.

According to author and environmental/legal scholar Professor Carl Cranor, there are approximately one hundred thousand manufactured substances or their derivatives registered and in common use by corporations in the United States alone. Of this "universe" of substances, environmental scientists know little. In the only major research regimen of its kind the National Academy of Sciences produced a 1984 study that identified more than 12,860 substances produced in volumes greater than a million pounds. Of those large-volume chemicals used in the United States, absolutely no toxicity information was available for nearly 80 percent, according to Cranor.

For the thirteen thousand chemicals manufactured in amounts less than one million pounds, three quarters had no toxicity data. Of the 1,800 drugs that the National Academy of Science reviewed, 25 percent had no toxicity data. Approximately 1,600 cosmetics had no toxicity information available. Of the 3,300 pesticides reviewed—all with known biocidal characteristics—almost 1,000 had no toxicity data. And perhaps most astounding, of the 8,600 food additives studied by the researchers, 46 percent were without toxicity data—yet approved for human consumption.

More than a decade after the National Academy of Sciences study, the Environmental Defense Fund—a New York–based "eco think tank"—published a follow-up to the National Academy of Sciences study. Of one hundred (randomly selected) high-volume chemicals that they reviewed, 71 percent didn't meet the minimum data requirements for health screening. Reproductive toxicity tests had not been done on 53 percent. Carcinogenicity tests hadn't been conducted on 63 percent. Immunotoxicity tests were not performed on nearly 90 percent of the chemicals.

Ecological risk assessment is not only inordinately cumbersome but also almost wholly ineffective at the critical task of determining a chemical's potential for harm—and it has failed in the broadest sense. In the two decades or more of its dominance as the foremost method of analyzing the safety of industrial chemicals, tens of thousands of substances were allowed to be commercially developed and used without any true understanding of their human health risks or ecological impact because risk assessment was not adequate to the task. Chemical risk assessment was—still is—a travesty by any reasonable metric. We've learned a bit about what has happened, so let us now look at how and why we got saddled with risk assessment. . . .

In its 2001 report, the European Environment Agency, discussing the plague of industrial contamination confronting the agency's member nations, wrote, "The costs of preventive actions are usually tangible, clearly allocated and often short term, whereas the costs of failing to act are less tangible, less clearly distributed and usually longer term, posing particular problems of governance. Weighing the overall pros and cons of action, or inaction, is therefore very difficult, involving ethical as well as economic considerations."

While presented a bit bureaucratically, the problem is nicely delineated by the above. Perhaps we could state things more directly by saying that Western industrial culture has always given far greater weight to economic considerations than to environmental ones, and regulation of industrial chemicals is a reflection of that bias. In any case, let's contemplate a hypothetical example of how this philosophical bias might play out in real life.

Big Oil Inc. has developed a new chemical that it wants to add to gasoline and sell at gas stations. The additive, they claim, has taken ten years and $50 million in research and development to produce. (Of course all such figures are crunched by the corporation and can be exaggerated outlandishly.) Big Oil Inc. contacts the ARA—the Appropriate Regulatory Agency—and they enter into a typical "permitting" dance.

The ARA regulators ask Big Oil Inc. how dangerous the product might be for human health and the environment and if they've done "risk assessment." Big Oil Inc. replies that yes, they certainly have and they find their product is harmless both to people and to the environment and they will be glad to pass along their voluminous data, which they do. Deep inside ARA a sharp-eyed toxicologist reviews Big Oil Inc.'s risk assessment data and sees that one of the chemical components of the gas additive has been shown to be an endocrine-disrupting chemical.

Since the plan is to infuse billions of gallons of gasoline with the additive, the toxicologist warns his superiors. They in turn contact Big Oil Inc. and ask them if another chemical might be substituted for the known endocrine disrupter. Big Oil Inc. responds, "not a chance" and furthermore they reply that if they are forced to remove the product, not only will they lose their $50 million investment, they will lose tens of millions of dollars more in projected revenue, a thousand new jobs manufacturing the additive won't be created, *and* the national effort to become energy independent will be grievously hurt. Besides that they tell ARA that their scientists do not deem the research on the chemical being an endocrine disrupter as "sound science" since there was no certainty that the chemical did any harm to humans, because all the previous research they could find was done with lab animals.

One can easily imagine the outcome of our parable—the chemical remained in the additive and the toxicologist got a new position at an ARA facility in the Aleutians. An epidemiologist or statistician might label this scenario as the weighting of "false negatives" over "false positives." And while a bit arcane, it is an important concept to appreciate as it informs the decision-making processes involved in risk assessment. A false negative occurs when an industrial chemical is regarded as not harmful when it actually is harmful. In other words, when analyses were completed, they resulted in a "negative" outcome for harm—a no-harm result. There are many examples of disastrous false negatives in the history of synthetic chemical exposures, with perhaps the preeminent being PCBs. (Some other examples of false negatives are DES, CFCs, benzene, and tributyl tin. Examples of nonchemical false negatives would be asbestos and overuse of medical X-ray technology.)

A "false positive" occurs when an industrial chemical is regarded as harmful and banned or restricted, and later research finds it to be benign. In his monograph for *Environmental Health Perspectives* in 2006, David Gee, a European scientist researching the application of the precautionary principle, could not find a *single example* of a false positive for any industrial chemical after a thorough search of the relevant literature. At first blush, this seems surprising. However it shouldn't be. As we've seen, the system, biased by corporate influence, is designed to be far less protective of public health than economic interests.

While the precautionary principle is clearly a concept whose time has come, given the abysmal record of risk assessment in protecting people and the environment from industrial chemical scourges like PCBs, it is anything but inevitable that it will be put into actual practice in Europe, let alone the United States.

Leading the opposition in the Bush administration was a fortysomething ex–Harvard professor, John D. Graham. Graham, an ultra-powerful but little-known industry advocate within the upper-level Bush executive team, felt that the precautionary principle was a concept that was as amorphous as it was dangerous.

Speaking for the "United States government" at the Heritage Foundation in 1996, Graham stated that "notwithstanding the rhetoric of our European colleagues, there is no such thing as the Precautionary Principle." Graham went on to explain that the precautionary principle didn't exist because there were nineteen different versions in "various treaties, laws and academic writings." It was an interesting philosophical position, according to critics of Graham. As one conservationist put it, "Isn't that like saying that trees don't exist just because there are a variety of definitions of what a tree is?"

Oddly, after his claims that the precautionary principle didn't exist, Graham still felt the need to attack the dangerous nonexistent concept by opining that the precautionary principle was not needed because the hazards that it dealt with *didn't exist either.*

Graham seemed to feel that the industrial chemical hazards documented by thousands of scientific experiments were, for the most part, little more than the fulminations of scaremongering researchers working at the behest of radical environmentalists. Once characterizing the concern over PCBs as "flustered hypochondria," Graham went on to list the hazards that in his opinion were phantoms created by anti-business forces: "Consider the following scares: electric power lines and childhood leukemia, silicone breast implants and autoimmune disorders, cell phones and brain cancer, and disruption of the body's endocrine system from multiple, low-dose exposures to industrial chemicals. In each of these cases, early studies that suggested danger were not replicated in subsequent studies by qualified scientists."

Then as a finale to his speech, Graham sketched the two major "perils" that being "too cautious" presented. The first and most important was that innovation would be "stifled." Graham charged that if, under the precautionary principle, "no innovation shall be approved for use until it is proven safe" then he opined that we'd be living as primitives, without "electricity, the internal combustion engine, plastics, pharmaceuticals, the Internet, and the cell phones."

For his second "peril," Graham took on the mantle of a dedicated environmentalist, saying that while this peril was "more subtle" it was perhaps more dangerous: "Public health and the environment will be harmed as the energies of regulators and the regulated community are diverted from known or plausible hazards to speculative and ill-founded ones." In other words, if regulators and the regulated got too caught up in worrying about issues like industrial chemical pollution and the precautionary principle, they wouldn't have time to think about problems that Graham believed were truly pressing—such as getting children to wear safety helmets when bike riding.

Unsurprisingly, the problem of serious conflict of interest came to the fore in Graham's confirmation hearing when Joan Claybrook, a longtime citizens' advocate, entered the following into the record:

> Over the past decade, the Harvard Center for Risk assessment (HCRA) directed by Graham has received unrestricted funding from 100 major industrial corporations and corporate trade associations, including oil, energy, chemical, agribusiness, mining and auto interests, such as Monsanto, National Steel, Kraft Foods (a subsidiary of Philip Morris), Atlantic Richfield, Ford Motor Company, Dow, 3M, DuPont, Exxon, the Chlorine Chemistry Council, the American Automobile Manufacturers Association, the American Petroleum Institute, the American Crop Protection Association, and the Chemical Manufacturers Association, now called the American Chemistry Council.

While John Graham may have been the most able of the Bush II anti–precautionary principle forces, with his superior academic training and Harvard pedigree, he wasn't the highest administration leader working on the issue. That distinction would be shared by Ann Veneman, secretary of agriculture, and Colin Powell, secretary of state, both of whom worked hard to undermine the EU adoption of the precautionary principle.

There was, though, a lone voice at the cabinet level defending the precautionary principle—that of Christine Whitman. Just prior to her confirmation hearings, Whitman took a bold stance in a speech to the

National Academy of Sciences that was perhaps the first acknowledge-
ment by a U.S. government official that the precautionary principle
even existed. Whitman stated her position for the record, eloquently
and succinctly:

> I believe policymakers need to take a precautionary approach to
> environmental protection. By this I mean we must 1) acknowledge
> that uncertainty is inherent in managing natural resources, 2) rec-
> ognize it is usually easier to prevent environmental damage than
> to repair it later, and 3) shift the burden of proof away from those
> advocating protection toward those proposing an action that may
> be harmful.

Of course, we've seen that Whitman's career as head of the EPA was
abbreviated by her stance on PCBs, General Electric, and the Hudson River, but her third point above—the need to "shift the burden of
proof"—may have hastened her departure, at least if the industry operatives who infested all levels of the Bush II White House took notice.

What is sure, though, is that the very heart and soul of the precautionary principle lies in shifting the burden of proof. Even to the most
detached citizen, it is common sense that the burden for proving a substance isn't harmful should be borne by whoever wants to disperse it into
the biosphere. And this simple, rational position, completely aligned
with common sense and basic ethical responsibilities, seems to strike
the deepest insecurities into the minds of the chemical industry leadership and their paid minions in the scientific community.

In a polemic that could be viewed as the mainstream chemical industry take on the precautionary principle, editorialists for the journal
Nature Biotechnology in 2001 dubbed it "a wolf in sheep's clothing" promoted by a "consortium of radical groups." The editorial writers went
on to tartly express their opinion that the rise of the precautionary
principle was "really about how a small, vocal, violent group of radicals wants to dictate to the rest of us how we should live our lives."
They concluded with a resounding plea for protection of beleaguered
corporate interests against the democratic environmentalist rabble:

"Bullies should not be permitted to use untruths, conspiracy, and violence to oppose legitimate research into technologies that can improve our safety and well-being. We should no longer allow extremists to dictate the terms of the debate."

Will the new Obama administration embrace toxics reform and the precautionary principle? Will President Barack Obama's appointees provide the muscle for carrying out the diffuse eco-reform promises made so often during his presidential campaign? It seems unlikely given Obama's predilection for avoiding controversy and taking a position of "benign neglect" toward the behavior of many corporate entities in the first year of his administration.

Actions speak louder than words, goes the old saw. In the case of new EPA administrator Lisa Jackson, she has called publicly for adoption of the precautionary principle as a legal bulwark underpinning any new environmental legislation aimed at controlling toxic chemical pollution. However, when asked—beseeched—by the Lone Star Chapter of the Sierra Club to review the dumping of Hudson River PCBs in West Texas, Jackson wouldn't step in to temporarily halt the rail shipments.

To mouth soothing phrases and tout laudable, sometime-in-the-future campaigns for taking a precautionary approach to toxic chemical contamination may be good for lulling a benumbed public into believing that the "audacity of hope" lives in the bosom of the Obama administration, but it clearly cannot mitigate the threat posed by a single PCB molecule. The dumping of Hudson River PCBs in West Texas was seen by some in the environmental community as a crucial initial litmus test of Jackson's resolve to truly deal with the realities of toxic chemical pollution and the abject failures of the Bush administration. But the new EPA administrator refused to stop the massive project—even though the very agency she headed had violated the spirit of the law by not filing environmental impact statements on the greatest transfer of toxic materials in the nation's history.

CHAPTER 17

Epigenetics, PCBs, and Us

Epigenetics is the next layer of complexity beyond the DNA sequence.

MATTHEW D. ANWAY, *Journal of Endocrinology*, 2006

The sequencing of the human genome will go down in scientific history as a magnificent achievement, no doubt. But the mapping of our genes may eventually be seen less as a great accomplishment than as an entrée into a new world—a picking of the lock on an immense vault containing untold sets of perplexing, intricate laws and biological interfaces.

The sequencing of the human genome captures what scientists call "the blueprint of life"—the genetic code contained within every cell of our bodies. But our DNA contains infinitely more hidden information than geneticists realized as they laboriously sequenced the genome in the 1990s. How did scientists come to understand that there is much more data encoded in our DNA than gene sequencing shows? The answer is simple and profound—and distinctly illustrated by the lives of identical twins. Identical twins have precisely the same DNA, but there can be wide differences in how their genes express themselves. For example, one identical twin will develop normally and the other might be severely autistic. Scientists cannot explain, on the basis of pure genetics, why there can be such radical differences between twins with identical DNA. As one advanced researcher puts it, simply enough, "something else must be at play." What might that "something else" be?

Actually it's many things. The study of how our genetic blueprint is modified by our environment is the study of epigenetics. In other terms, it's the study of the "nurture" of our genes, if genomics is the study of our genes' "nature." While this might make epigenetics sound like a standard biological process, it is anything but. The brutal Darwinian model of adaptation via natural selection, followed by Mendel's founding of genetic science based on the dynamics of mutation, essentially joined natural selection with genetic inheritance as the twin engines that determined the form of life on Earth, making Darwin's and Mendel's discoveries the two modern pillars of biological scientific dogma.

But as of the new millennium, scientists began seeing the first hazy outlines of a third pillar, as powerful in determining the quantity and quality of life on the planet as those conjured by Darwin and Mendel. In 2000, two researchers at Washington State University, Matthew D. Anway and Michael K. Skinner, began intensively studying epigenetic impacts by dosing pregnant female rats with a known endocrine-disrupting fungicide, vinclozolin. (Vinclozolin is widely used in the United States to control several types of fungi in grapes, strawberries, and other fruit, as well as vegetables and ornamentals such as rhododendrons and azaleas. It has been banned in Denmark.)

Their work, like that of Soren Jensen, the discoverer of PCBs in the biosphere, would startle and absorb biologists, chemists, toxicologists, and geneticists across the globe. The first generation of male rat pups whose mothers received doses of the fungicide turned out to have fertility problems as well as a number of other diseases that developed later in their life cycle. Certainly this was not a major finding since we've already seen the damage that PCBs—a "kissing cousin" of organochlorine—can do to the offspring of women exposed during gestation.

But what came next in the experiment with the fungicide may well change the direction of all biological sciences—including human medicine—in ways so radical that it is impossible to foresee. Anway and Skinner found that the changes that the fungicide wrought in the rats were *transgenerational.*

The disease states initiated by exposing the mother rat to the fungicide not only affected the first generation of males that she produced, but the second, third, and fourth generations also suffered from the same types of diseases. In human terms, what Anway and Skinner had discovered is that—theoretically at least—if your mother was exposed to certain endocrine-disrupting chemicals during pregnancy, not only might you be affected, but your children, your grandchildren, and your great-grandchildren could be affected too.

To be sure, Anway and Skinner were up front about dosing the pregnant rats with much higher levels of vinclozolin fungicide than anyone would be likely to encounter in a glass or even a bottle of wine. But that wasn't the point.

The point was, Anway and Skinner had demonstrated that a synthetic chemical, a commonly used fungicide, had the ability to change how genes did their job—how they expressed themselves across multiple generations—by only a single exposure of the matriarch, *without any observable change in the DNA sequence of the test animals.*

When dealing with endocrine disruption, the impact on organisms such as mammals is usually quite subtle. But the spooky transgenerational effect that Anway and Skinner discovered was hardly subtle. Of the offspring that were bred over four generations, approximately 85 percent developed serious conditions—breast tumors, prostate disease, kidney disease, immune system abnormalities, and premature aging. Some of the mice developed only a single disease, but most had inherited multiple ailments through four successive generations.

Since the experiments were highly complex, specialized, and expensive to execute, Anway and Skinner's transgenerational epigenetic effects have not yet been replicated, as of this writing. But the transmission of epigenetic mechanisms in a single generation, rather than transgenerationally, has been documented in a series of already famous experiments using agouti mice by Dr. Randy Jirtle of Duke University.

When pregnant mice were exposed to BPA (bisphenol A), a common chemical contaminant found in many types of plastic containers and known to have strong endocrine-disrupting capabilities, their offspring were obese and had significantly higher rates of cancer and diabetes when they matured. Jirtle could also change the fur color of the offspring by dosing pregnant mice with BPA, in a literally eye-catching demonstration of epigenetic effects—since fur coloration, or hair color in humans, for that matter, was long thought to be determined solely by genetic inheritance.

While changing the appearance of the mice through epigenetic manipulation might have been a sort of arresting scientific parlor trick, the rates of obesity induced in the mice epigenetically was a much more interesting and disturbing discovery. It led researchers like Jirtle to speculate that given the widespread exposure of women in the United States to BPA during pregnancy, the rogue chemical might be contributing to the epidemic of obesity in the current generation of American children. "There could be a connection between the increase in plastics in our environment and the rising incidence of obesity in humans," Jirtle told PBS in the fall of 2007. "However, such an association will not be able to be demonstrated unequivocally until the expression and function of genes involved in human obesity are shown to be altered by BPA." (It should be noted that human body burdens of BPA in the United States sometime exceed those of PCBs and DDE. BPA has been linked to a number of disorders in test animals, including early puberty, breast cancer, diabetes, and low sperm count.)

Other studies, this time of human beings, found that epigenetic effects occur not just in the womb, but apparently throughout the human life span. A Spanish researcher, Dr. Manel Esteller, and his associates at the Cancer Epigenetics Laboratory in Madrid studied more than three dozen pairs of identical twins, ranging in age from toddlers to the elderly. What he uncovered was an eye-opening trend in his subjects.

The younger sets of twins, who had lived together for the most part, had gene expression patterns that were, for all intents, identical. But the more mature twins, who had developed different lifestyles and had lived in geographically disparate locations, had dissimilar patterns of development in the tissues Dr. Esteller studied. Esteller discovered that a pair of fifty-year-old twins had 400 percent more genes that were expressed differently than a set of three-year-old twins in various tissue sets, such as mouth cells, abdominal fat, and certain muscle groups. In other words, due to epigenetic changes, the identical twins were no longer identical even though their genes were still exactly the same.

What were the mechanisms that gave epigenesis its profound ability to apparently alter all forms of higher life? The answer, even at such an early point in the scientific unraveling, is complicated and controversial, but there are points of agreement that provide some solid footing for researchers. It seems that epigenetic processes take a variety of forms at molecular levels of activity. However they all are linked in their ability to turn off or turn on specific genes within a cell. In order to understand the importance of this switch-throwing capability, we must first understand the role played by this action.

Every cell in our body has a complete set of the same DNA (strands of gene sequences) found in every other cell—the combined codes we inherited from our mother and our father. So one might ask, "How is it that we have eyes, ears, lungs, livers, hearts, and brains when all our cells have the same set of blueprints?" The answer, in the simplest terms, is that there are gene "switches" that turn cells on or off, allowing them to differentiate without changing our DNA sequences.

Early in the development of mammals and other animal life, a few key genes are tasked with being "switches," commanding cells how to grow and replicate within different regions of the organism. Later in the development of the animal, epigenetic mechanisms begin performing the crucial duty of switching genes within the cells on and off. How this is done—how epigenesis works—is, as we've pointed out earlier, one of the newest frontiers of biological research.

Scientists like Jirtle, Anway, and Skinner have established beyond doubt that epigenetic changes occur. But determining the mechanisms used is another matter, and the subject of intensive study involving exquisitely complex cellular processes such as methyl tagging of gene base pairs, the folding of DNA, imprinting, reprogramming, and transcription of genetic information—none of which we shall delve into save to reiterate that the processes are real, even while the mechanisms are not yet anywhere near being fully understood.

The spectrum of disease that epigenetic changes are linked to is broad, to say the least. For almost every cancer group that has been studied, there seems to be some evidence of association with epigenetic alterations. The latest research brings us this abridged list by scientists who have studied epigenetically induced illness: cognitive dysfunction, respiratory and cardiovascular disease, reproductive, autoimmune, and neurobehavioral illnesses.

The variety of suspected catalysts for epigenetically induced disease is also wide and getting wider. It includes the entire spectrum of endocrine disrupters: heavy metals, pesticides, diesel exhaust, tobacco smoke, and most types of aromatic hydrocarbons, as well as naturally occurring hormones, viruses, bacteria, and even some basic nutrients, as demonstrated by Jirtle with his agouti mice.

With this we return to Ted Swann's "magic fluid," since PCBs are among the most powerful synthetic endocrine disrupters ever discovered. In a forceful study, Dr. David Crews, a current leader in epigenetic research, established beyond any doubt the power of PCBs to modify gene expression in the most basic way—by altering sex. In the early 1990s, Crews and his colleagues Judith Bergeron and John McLachlan performed a series of stylish experiments using eggs from the common red-eared slider turtle. For many species of turtles—like their reptile relatives, alligators—the sex of offspring is controlled by the temperature levels that the eggs experience within the nest. Warm incubation temperatures beget female turtles and colder incubation temperatures produce males.

Applying PCB congeners known to have endocrine-disrupting abilities, Dr. Crews and his associates found that eggs coated with minute amounts of the chemical (low parts per billion) produced females at temperatures that would usually produce all males. Astonishingly, in some of the experimental egg groups, coating the eggs with specific PCB congeners produced 100 percent female turtles when all should have been male—thus *totally* reversing the sex of the baby turtles. Crews also confirmed Fred vom Saal's low-dose-effects research for endocrine disrupters, finding that he could reverse the sex of 14 percent of the turtles by dosing them with only forty parts per trillion of PCBs. (Recall that a part per trillion would be equal to the width of a hair compared to the circumference of Earth.)

Given the results of the experiments, Dr. Crews reasonably speculated that PCBs could well be altering the sex ratios of entire wildlife populations. Bergeron, Crews, and McLachlan noted without editorializing that the amounts of the chemical used on some of the baby turtles were approximately the same as average levels found in human breast milk in industrialized countries. They ended their seminal paper by explaining that, given the easily replicable results of their turtle study, they could only conclude that there was no safe "threshold" dose for PCBs and hundreds of other fellow-traveling industrial chemicals.

Their experiments clearly showed that during critical times of development, infinitesimal amounts of polychlorinated biphenyl could radically alter the growth of an organism and such fundamental characteristics as its sex via endocrine system disruption. And these severe disruptions occurred at levels of exposure always thought to be safe. The scientists closed their paper with an ominous message veiled in low-key, minimalist terms: "Therefore, we expect our conclusions are not idiosyncratic, but rather are widely applicable. Our findings and their implications reinforce our concern for the health of humans and wildlife exposed to these low doses."

Given the characteristics of PCBs shown by Bergeron, Crews, and McLachlan in their seminal turtle experiments, there is little doubt in the mind of epigenetic researchers such as Michael Skinner that PCBs are extremely powerful epigenetic agents. Making them even more dangerous than fungicides such as vinclozolin, is the fact that PCBs, even after thirty years of banishment, are still—in many parts of the globe—the most prevalent group of chemical contaminants. But it is unlikely that PCBs will be studied in depth by researchers probing the new field of epigenetics, even with the threat that they present to the normal expression of Earth's gene pools.

Professor Crews, long a student of PCBs and the politics surrounding Swann's "magic fluid," ended his interview with me on a deeply pessimistic note. Knowing that two thirds of all the PCBs produced are still waiting to invade the biosphere, he thinks the only solution to the problem of industrial contamination of the biosphere by PCBs and other rogue chemicals will be "end of pipe" solutions—remedial efforts that deal with offsetting their endocrine-disrupting effects *after* they've been absorbed by humans or other animal life.

As far as getting needed funding for epigenetic research on PCBs approved, Crews is not sanguine. "The problem with getting PCB research money is that PCBs have been banned. Our current politicians can point to that and say that PCBs are not an issue anymore. Besides, they weren't around when PCBs started becoming a problem . . . and you *know* that they don't like to deal with other people's mistakes."

Clouds and Sunlight

We are now finding that compounds like the chlorinated and
brominated flame retardants, for instance, which are used widely in the U.S.
and the UK and throughout the world, resemble the structure of the PCBs.
So we're working on comparisons between old and new structures, old and new uses,
and we're finding that we are repeating the same problems with these new
chemicals we encountered with the old chemicals.

DR. BRAM BROUWER, MAY 2004

The Medina Electric Cooperative is a small electrical utility that serves the Hill Country of Texas. Established during the Depression by President Franklin D. Roosevelt, it is a well-run outfit with a proud heritage. FDR—with cheerleading by a young congressman named Lyndon Baines Johnson—had made a second-term campaign promise to the hardscrabble ranchers living in the remote and lovely hills of central Texas that there would be light—and the president kept his word. With the passage of the Rural Electrification Act, small electric cooperatives sprouted across America and thus Medina Electric was born—the literal bringer of light and a better life to the Texas Hill Country.

Today the Medina Electric Cooperative is a modern utility that offers all the amenities of the Internet age while still retaining a casual, mellow approach in dealing with its mostly rural customer base—of which this author is one. When I originally called to ask their one-person public affairs department about how they were handling PCBs, I was put in contact with no less than the chief engineer of Medina Electric. Genially wary, he told me that their electrical grid did indeed have transformers with PCBs and that they handled them on a "case by case" basis—a bureaucratic locution for not having any real plan.

Three years later, while writing this chapter, I called Medina Electric back to see if they had decided to take PCBs more seriously—and indeed they had. Medina Electric's CEO, Mark Rollans, told me that there were no longer any transformers that he knew of that contained PCB-laced transformer oil. A parent organization, the Texas Electric Cooperative, a group of smaller utilities all essentially with the same provenance as Medina Electric's, had disposed of the PCBs in Medina's power grid using EPA-approved handlers who had incinerated the PCBs according to government standards.

(Perhaps you are wondering, as I did, just what Medina Electric used as the replacement for PCBs—a substitution that Monsanto, General Electric, and even an EPA administrator warned could cause massive power outages across the United States and the globe? According to Rollans, their replacement for the "magic fluid" was . . . plain old mineral oil.)

At the end of our interview, Rollans said he'd like to add something, not as a spokesman for the utility, but as just a regular co-op member—just another customer living in the Hill Country. He said that dealing with PCBs was difficult and expensive for Medina Electric, but what angered him most was that they received no assistance in handling the PCBs that General Electric had been selling them for years. "They just walked away from the problem," he said. "They wiped their hands of it, never offered any help."

Probably by the time the folks at Medina Electric realized what had been foisted off on them by GE, it was too late to sue. More likely, though, the independent-minded Texans who ran Medina Electric wouldn't have taken GE to court in any case. That just wasn't the old Texas way. And they certainly weren't about to go around to customers with transformers on their property and make them sign "hold harmless" agreements. Medina Electric just bit the bullet and dealt with it.

Nobody knows the numbers, but American utilities, big and small, public, private, and cooperative, are still stuck with tens of thousands of transformers containing untold tons of PCBs that will inevitably be released into the environment as the transformers age and leak or are knocked out by natural disasters such as Hurricane Katrina—unless they are individually decommissioned. Since there are no federal laws that specifically require that PCBs be removed from "closed systems" such as transformers, electrical utilities react to the potential environmental disaster that PCBs represent in different ways. Some, like California's Pacific Gas and Electric, dealt with PCBs starting in the late 1980s. By the mid 1990s, PG&E claimed that it had removed all PCBs from its power grid infrastructure. Others, especially the smaller electrical utilities such as Medina Electric, moved more slowly. And some have moved hardly at all.

Just how many pounds are out there nationwide in transformers made prior to 1979, when Monsanto supposedly stopped selling PCBs, is anybody's guess. Transformer manufacturers General Electric and Westinghouse, the two largest suppliers of electrical grid equipment to utilities and industry, surely have good, hard data on the amounts of polychlorinated biphenyl they put into transformers over the years, but they haven't made those figures public, nor is it likely that they will.

While PCBs in electrical transformers are the big-ticket item to worry about, they don't represent the greatest threat to the environment from a quantitative standpoint. That title is held by another ubiquitous, but much smaller piece of electrical gear: the capacitor. Capacitors are basically energy storage devices that assist electrical circuits in performing whatever their designated duty is. Manufactured mainly by Westinghouse and General Electric in the United States, PCB-filled capacitors were used in a wide variety of appliances: air conditioners, copy machines, dehumidifiers, early microwave ovens, portable oil-filled heaters, mercury vapor lamps, and most commonly, in fluorescent light fixtures. According to United Nations statistics taken from the last comprehensive PCB inventory in the United States, there were 630 million pounds of PCBs being used in capacitors. This is almost double the amount believed to be in transformers (335 million pounds) still in use at the time of the survey.

Regardless of how they were used, there are basically two methods for disposal of PCBs found in electrical equipment or hydraulics systems. If the PCBs are in relatively concentrated form, they can be incinerated. But because of their unique chemical ability to resist heat and breakdown, incinerators must reach temperatures in excess of 2,000 degrees Fahrenheit in order to fully combust PCBs. Consequently, it is an expensive and highly demanding process.

Incineration is also a relatively new practice. Traditionally, PCBs have been discarded in municipal landfills or dumps since they were developed in Germany over a century ago. There are no estimates—reliable or otherwise—for how many thousands of tons of PCBs reside in dumps and landfills. What is known is that unless the landfill is state of the art, PCBs won't stay put. They leach into groundwater and nearby bodies of water or they volatilize into the atmosphere. Once in the air, these airborne PCBs are a vast source of global pollution, contaminating even the most pristine and remote regions of the planet.

However, well-designed landfills created especially for handling large volumes of PCBs, such as those dredged from the Hudson River, can minimize evaporation and can generally avert leaching. While experts believe that environmentally sound landfills should be considered only temporary solutions for disposing of PCBs, like the storage of nuclear wastes, they do appear to offer excellent potential for a long-term resolution. Perhaps the most promising strategy would be biodegrading landfill-held PCBs using "designer" bacteria that have been genetically modified to "eat" the contaminant as well as by-products.

Another "end of pipe" possibility for dealing with PCBs captured in landfills or circulating anywhere in the biosphere, as envisioned by computer genius and futurist Ray Kurzweil, may be through using "nanobots"—molecular-level robots measured in microns, or millionths of a meter. As he writes in his controversially optimistic book *The Singularity Is Near*, "Once we can go beyond simple nanoparticles and nanolayers and create more complex systems through precisely controlled molecular nanoassembly, we will be in a position to create massive numbers of tiny intelligent devices capable of carrying out relatively complex tasks. Cleaning up the environment will certainly be one of those missions." Even more fantastic, Kurzweil predicts that within fifty years not only will nanobots be cleaning up PCBs in the biosphere, but we will have PCB-detoxifying nanobots coursing through our bloodstreams. (Current nanobotic research seems to be proving Kurzweil's visions are not just pie-in-the-sky stuff. Citing a study published in 2001, the futurist writes that experiments have already established that nanobots can more effectively absorb industrial contaminants like "PCBs, pesticides and halogenated organic solvents" than traditional methods that rely on activated carbon.)

Regardless of coming scenarios, the threat that PCBs pose to human beings and our ark mates can be reduced right now *if* the proper steps are taken—and taken quickly. Scientists have now established, beyond doubt, that careful removal of PCBs works to safeguard the environment. Studying wildlife at a decontaminated military site at Saglek Bay, Canada, investigators found stunning reductions of PCB burdens in fish and fish-eating birds within eight years of the decontamination. After the removal of twenty thousand cubic meters of PCB-saturated soil, body burdens of PCBs in local wildlife were approaching the levels where no effects on reproduction can be measured. PCB levels in sediments, fish, and birds "decreased by 10, 19, and 6 times, respectively," according to the Saglek Bay researchers.

What this landmark study clearly shows is that cleanup efforts that involve the removal of the source of PCBs can really work. And they can work rapidly. Critical to such efforts, of course, is approved disposal through incineration or state-of-the-art landfills, but additional actions need to be undertaken as well. All foods need to be monitored for PCB content. This is particularly important for fish and dairy products, which readily accumulate them. Strong warnings must be issued as to the dangers of eating foods with even comparatively low levels of PCB contamination, especially for women of childbearing age. In addition, outreach programs need to be started and maintained in every community where economically disadvantaged people fish for food.

As far as individuals are concerned, although there has been very little research on the subject, experts believe that taking antioxidants such as vitamin C or vitamin E can possibly limit the damaging cellular-level effects of PCBs. Experiments at the University of Kentucky showed that when cells were simultaneously exposed to vitamin E and PCBs, the vitamin was able to block cellular damage. The results were significant enough for one of the investigators to recommend vitamin E supplements for at-risk people such as those who clean up PCB-contaminated sites.

Whatever precautions are taken, it may not be enough for many of us. The weight of evidence of the most recent and best research clearly indicates that PCBs will probably continue to affect the health of tens of millions—perhaps hundreds of millions—of people who are genetically susceptible to diseases linked to PCBs. As we've seen, these increased risks include everything from relatively benign autoimmune diseases in infants to the catastrophic health problems that cancer victims must face. These wide and sometimes lethal risks that PCBs pose to a swath of human genotypes will remain as long as large amounts of the chemical are circulating in the biosphere—or until the dreams of Ray Kurzweil come to fruition.

Of course, such anthropocentric concerns don't take into consideration the silent destruction of animal life on the planet due to PCB toxicity, synergy, and ubiquity. But we're all sailing on the same ark, and if we save ourselves through a zealous international campaign to locate, identify, and dispose of the approximately one and a half million tons of PCBs that have been produced in the last seventy years, we will undoubtedly save at least some, if not many, of our ark mates who are at risk too.

While PCBs may make a good toxic "poster child" for a world bathed in rogue industrial chemicals, they are not necessarily the greatest future threat. In 1996, the last time an inventory was taken by the Environmental Protection Agency, over seventy-five thousand industrial chemicals were being produced by or imported into the United States. Let that "over seventy-five thousand" number roll around a bit in your mind. Then understand that it includes only chemicals manufactured in or imported into the United States in quantities *over* one million pounds. Manufacturing figures are supplied by the chemical makers themselves, such as Monsanto. Consequently, given their long histories of distorting data for their own interests, it must be assumed that the amounts they claim they produced are lowball figures at best. And *then* realize that these numbers don't even include the most important categories of chemicals, such as drugs, food additives, and pesticides.

The public knows almost nothing about the toxicity of the seventy-five thousand chemicals being produced and used by North American corporations, or what their impact on the environment may be. When researchers for the conservation group the Environmental Defense Fund did a random analysis of one hundred of the highest-volume chemicals in use, they found that nearly three quarters "did not meet the minimum data requirements for health hazard screening."

In reality, the screening of almost all the chemicals released into the biosphere is controlled by the very industries that manufacture them— creating the ultimate in conflicts of interest. The testing of industrial chemicals is usually done by private research labs, such as IBT, contracted by the manufacturer of the chemical. But according to Dr. Fred vom Saal, a world leader in the field of developmental biology, the data the contract lab gathers about the new chemical product isn't evaluated for toxicity by unbiased scientists. Instead, the toxicity data "goes directly to corporations through their legal departments," says vom Saal. "Then they decide whether to provide it to the government or not. They decide whether the outcomes are adverse. That could be very subjective. And they just say, 'Well, we didn't provide this information because we didn't think it was a problem.'"

If tens of thousands of chemicals are being released into the environment every year without anyone knowing their impact on human beings (or with the health effects known but suppressed by corporate legal counsel), one can assume that even less is known about their effects on the biosphere and global ecosystems, and less still about the synergistic characteristics of these industrial contaminants as they circulate together.

The chemical industry and its clients continue to baste the United

States and the planet with billions and billions of pounds of synthetic substances that are toxic, bioaccumulative, and often nearly indestructible by natural processes. And in contrast to Europe, where citizens are visibly and vocally outraged by industrial chemical contamination, in the United States, aside from some exceptional regional and local environmental reporting, there is nary a peep from the public or politicians. As one activist said regarding the nonreaction of the average citizen to the contamination of all edible fish in the Great Lakes by PCBs, "They ought to be protesting in the streets."

Even more disheartening is the public's silence about the well-documented contamination of fetuses and newborns with hundreds of toxic industrial chemicals besides PCBs—many as yet not even identified. According to a study completed in the summer of 2005 by the Environmental Working Group, the umbilical cords of ten randomly selected newborns were found to be contaminated with 287 "commercial chemicals, pesticides, and pollutants" (with PCBs constituting the leading contaminant found as far as quantities). And according to the EWG scientists, the newborns were probably infested with hundreds of other toxic chemicals for which they couldn't afford to run analyses.

Renowned research scientists Philippe Grandjean and Philip Landrigan, in their surpassing and unusually candid review of the industrial chemical pandemic that humanity faces, sum up the problem this way in their seminal *Lancet* article of November 2006:

> A pandemic of neurodevelopmental toxicity caused by industrial chemicals is, in theory, preventable. Testing of new chemicals before allowing them to be marketed is a highly efficient means to prevent toxicity, but has been required only in recent years. Of the thousands of chemicals used in commerce, fewer than half have been subjected to even token laboratory testing for toxicity testing.
>
> Nearly 3,000 of these substances are produced in quantities of about one million pounds every year, but for nearly half these high-volume chemicals no basic toxicity data are publicly available, and eighty percent have no information about developmen-

tal or pediatric toxicity. Although new chemicals must be tested more thoroughly, access to these data can be restricted, because they could be claimed to constitute confidential business information.

Now, perhaps you are wondering whether there are more PCB-type catastrophes in our future. Undoubtedly there are. And many may surpass PCBs in their biocidal impact on the planet. Here, in no particular order, are just a few of the hundreds of candidates to be the next "magic fluid":

- BPA (bisphenol A): One of the highest-volume industrial chemicals produced in the world, with over six billion pounds per year manufactured. Used in a wide variety of plastics and now ubiquitous everywhere on the planet, BPA has been shown to have serious endocrine-disrupting effects.
- PFOA (perfluorooctanoic acid): A chemical developed and sold by DuPont in nonstick cookware and other consumer products that is completely resistant to natural breakdown and thought to be a human carcinogen.
- PBDE (polybrominated diphenyl ethers): Used as flame retardants. They are surpassing PCBs and DDE as the most common human toxic contaminant in some areas of North America.
- Phthalates: A group of industrial compounds used in everything from wood-finishing products to lipsticks. They are implicated in a wide variety of endocrine-disruption-related disease including genital abnormalities and premature births.

But now for the surprisingly good news: the future can be as bright as we want it to be regarding the control of the budding toxic costars of PCBs. There are two solutions to the problem of the toxic contamination of the planet and they are both quite simple—with the caveat that *simple* does not mean *easy*.

The first part of the solution involves the precautionary principle. While the application of the precautionary principle will require complicated political arrangements, solving the global toxic contamination crisis by applying it is nothing more than the use of common sense. Theo Colborn and her coauthors, Diane Dumanoski and John Peterson Myers, make this point concisely in their book *Our Stolen Future:* "Presently new chemicals are considered innocent until proven guilty. This should be reversed. New chemicals should be assumed harmful until they have been thoroughly tested for all the kinds of harm we presently know about."

Chemical industries, by influencing political systems, have been able to suspend rationality on a global basis when it comes to toxic industrial chemical pollution. Currently, corporations can put almost any substance they want into the environment unless the public can prove far beyond a reasonable doubt that it is harmful. But if we made the precautionary principle the rule of law, the chemical industry would bear the responsibility of proving that its products are harmless. Such a solution is both easily grasped and logical, and, if correctly pursued, would be effective in preventing any new "magic fluids" from being sprung on the biosphere. How the precautionary principle could be codified is outlined succinctly by conservationists at the Environmental Working Group:

- Require chemical manufacturers to demonstrate conclusively that the chemicals they sell are safe for the entire population exposed, including children. In the absence of information on the risks of prenatal exposure, chemicals must be assumed to present greater risk to the developing baby in utero, and extra protections must be required.
- Require that the safety of related chemicals (such as all synthetic organochlorines) be assessed as a group. The presumption would be that these chemicals have higher toxicity unless manufacturers clearly prove otherwise.
- Grant the EPA clear and unencumbered authority to demand all studies needed to make a finding of safety and to enforce clear deadlines for study completion.
- Eliminate confidential business protection for all health, safety, and environmental information.

The second part of the solution is even less complicated. It involves the answer to dealing with the tens of thousands of chemicals now in use by industries and corporations across the globe that would not be affected by use of the precautionary principle. It simply is this: *immediately ban any commercially manufactured chemical if it or its metabolites are found in the umbilical cords of human infants.* What gives the makers and users of industrial chemicals the right to assault our children with synthetic contaminants, invading their bodies almost immediately after conception, perhaps physically and neurologically altering their lives—or possibly dooming them to face cancer and other catastrophic diseases as adults?

Certainly drastic measures must be taken, and taken swiftly. But there are also subtle, long-term cultural changes that need to be effected—"conceptual shifts," as eco-theorist Dr. John Peterson Myers puts it. In simplified terms, the necessary new consciousness should encompass the following principles:

- "The dose makes the poison" should give way to accepting that extremely low doses of some industrial chemicals can be toxic and serious health impacts can be caused by nothing more than background levels of substances such as PCBs.
- Industrial chemicals believed to be "safe," from a carcinogenic viewpoint, can in fact disrupt critical hormonal systems, leading to other types of severe disease.
- Immediate cause and effect is not a viable way to test for toxicity. Long latencies are common and fetal exposures to industrial chemicals can lead to disease many decades later.
- Examining the impact of individual industrial chemicals doesn't work in the real world. There are too many synergistic effects that can occur when substances interact with each other.
- Traditional toxicology cannot protect us. Subtle effects on the reproductive and immune systems as well as neurological, cognitive, and behavioral functioning must be taken into consideration—not just mutational changes, carcinogenicity, and simple cell death.
- The current focus on adult human health must be redirected

toward fetuses, infants, and children, since industrial chemical toxicity is often greatest at a particular stage of development, with periods of high growth during childhood being by far the most vulnerable time.

While PCBs represent the greatest inadvertent poisoning of the biosphere in the history of the planet, there is still much we can point to that allows for optimism. Clearly we can—with the right political will—ameliorate the problem to a large extent, if never totally conquer it. Humanity may not find itself living on a planet uncontaminated by PCBs and other toxic synthetics for millennia, perhaps many millennia. But there certainly is the possibility that technology will provide the "end of pipe" solutions hoped for by scientists like Professor David Crews and futurists like Ray Kurzweil, which will at least reduce the impact of PCBs and their industrial chemical cohorts on us and the biosphere.

If epigenetic research continues in its present direction, then dire as the currently recognized effects of PCBs and other chemical pollutants have been for us and our ark mates, researchers may well discover in the coming years that the extent of the catastrophe was only minimally understood at the time these words were written. The lessons gleaned from Swann's "magic fluid" must be applied by undertaking a global crash program for dealing with industrial chemical contamination. It could save us from damnation by our descendants, and theirs—and theirs. Are we so biocidally selfish to do otherwise?

Acknowledgments

For giving me the chance to write these words, I would like to say thank you to Dr. Harry Watkins, an old country surgeon who rejected my six-month, dead-man-walking cancer prognosis, then with consummate skill cut me up and refurbished me (for a little while at least) with the help of his longtime sidekick, Dr. John Shudde. And to Marilyn Unruh, hospice nurse extraordinaire, my soul-felt thanks for your amazing prayers and the prayers of your congregation. To the other angels at Uvalde Memorial Hospital and Hospice Care: bless you. You all saved my life—with a lot of help from "Upstairs," of course.

Some historical acknowledgments are in order, since *Biocidal* was brewed from my days in the Northwest: my thanks to the Northwest Office of the Sierra Club and its staffers in the 1970s, especially Donna Klemka for her support as a colleague and dear friend. To David Brewster, publisher of the *Seattle Weekly*, thanks for publishing that first article on PCBs in the Duwamish in 1976—even though you rejected it initially.

Now, as to the people who helped me with the writing of this book— well, I simply can't express enough gratitude. Thanks to Andrew Proctor and Sarah Heller of PEN, the Authors' League of America, and to

John Hyde of the Fund for Investigative Journalism for their kindness and financial assistance. Despite tremendous work schedules and heavy responsibilities, literally dozens of people in the scientific community generously shared their views and thoughts, as well as criticisms and compliments. What a pleasure it was to interact with so many absolutely brilliant and dedicated people—even if I didn't always agree with their views. My sincere appreciation to Dr. Harry Smith, Dr. David O. Carpenter, Dr. Devra Davis, Dr. Theo Colborn, Dr. Fred vom Saal, John Ansley, Sister Patricia Daly, Dr. Bruce Blumberg, Dr. Deb Swackhamer, Eric Francis, Dr. Joseph Cummins, Dr. Peter Ross, Donald Stewart. Dr. Kirsten Moysich, Karen Finney, Kim Nauer, Dr. Arjan Palstra, Dr. Adria Elskus, Anita Hjelm, Cynthia Benson, Dr. David Hunter, Dr. Francine Laden, Dr. Gordon Gribble, Dr. Deke Gunderson, Dr. Lynn Hubbard, Dr. Jennifer Keller, Dr. John Fenn, Dr. Jocelyn Leney, Barbara McAlpine, Marilyn Fike, Robert McClure, Denis Hayes, Dr. Steven Mims, Dr. Pat Ganey, Dr. Per Jonsson, Dr. Philippe Grandjean, Dr. Russ Hauser, Richard Pollak, Robin Herman, Dr. Paul South, Dr. Stephen Safe, Dr. Susan Schantz, Dr. Petra Wallberg, Dr. Yawei Zhang, Dr. Mary Wolff, the Honorable Ogden Reid, Lou Valdez, and Dr. Richard Gambitta.

As to individuals who assisted me directly with the manuscript, my first thanks must go to a legendary scientist and researcher, Robert Risebrough. It was Dr. Risebrough who first identified PCBs in the Western Hemisphere back in the late 1960s. His generosity in sharing his papers and writings from this landmark period of discovery, as well as his careful editing of large portions of the manuscript, gave me a sense of confidence that has buoyed me through the periods of doubt that any author goes through when writing a complex investigative work. He may not agree with some of the materials in this book, especially on the passages involving endocrine disruption, but nonetheless I cherish his assistance.

Also a character historically pivotal to the PCB story is Robert H. Boyle . . . superb writer and fisherman and supreme river rat . . . who is responsible—more than any other human being—for saving the Hudson River on its deathbed. In the mid-1960s, it was Bob Boyle who talked his editors at *Sports Illustrated* into popping for extensive lab testing of

Hudson River fish (caught by Boyle himself), which turned up for the first time the extensive contamination of that gorgeous riverine environment. Boyle, a true living legend, was one of the very first Americans to recognize the threat of PCBs and ring the alarm bell—and he's still ringing it without signs of slowing down.

And finally, to the many midwives and mechanics who put their skills to work in getting *Biocidal* published: to John Oakes, the charming, astute editor/publisher; your never-flagging support and efforts were such a blessing and rarity—a wonderful counterpoint to the rudeness, superficiality, avarice, myopia, predatory practices, and ethical laxity that fester in too many of our American publishing houses and literary agencies these days. With appreciation and gratitude for all the hardworking, keen-eyed strategists and bibliophiles at Beacon Press: Susan Lumenello, Pam MacColl, Tom Hallock, Sarah Laxton, and the wonderfully industrious Reshma Melwani. And, of course, Alexis Rizzuto, my brainy, energetic, simpatico, and visionary editor at Beacon Press, who has a preternatural feel for the scientific and political highways and byways of PCBs—thank you.

And now to Writers House: working with the folks there has been the most rewarding professional experience in my thirty-five-year writing career. Maya Rock, who discovered *Biocidal*, is not only scary-smart; she is an organizational whiz, tasteful and discreet. Albert Zuckerman, my agent, is one of those rare individuals that one may or may not run across in a lifetime. To have such a talented and decent human being as your literary agent is, to my mind, akin to a Divine blessing. I won't go into the lengths that Al went to in order to get *Biocidal* correctly published, but I will say that "above and beyond the call of duty" hardly does justice to his efforts. Al, I salute you.

And to my exceptional friends, Greg and Patty Pasztor—my alterfamily—thanks for your warmth and caring. To my "dutch" brother, writer/photographer Joel Rogers, from his striking photographs to his ability to find places of peace and great beauty, an artist whose lifelong commitment to the environment is inspirational. And of course to my *favorite* daughter, Erin, and her precious brood, Lilah and Otis—always my love. But most of all to the Gracious One and the other presiding Gods and Angels: *merci!*

Notes

In writing *Biocidal* I worked with approximately 650 to 700 documents of various kinds, not including personal interviews. I wrote the manuscript from a 460-page hypertext-linked outline that took me from an outline entry to a primary document with a single keystroke. I have used my chapter outlines with their links to re-create chapter notes rather than doing them in the more traditional cite-while-you-write manner. The following are extremely condensed versions of sourcing for each chapter, a sort of "cherry-picking" of the most crucial information used in *Biocidal*. It allows any interested readers to get a feel for approach and execution. The provenance for all direct quotes is specified in the text.

CHAPTER 1 — THE MAN WHO POISONED THE PLANET

A good percentage of the material contained in chapter 1 comes from *The Triumphs and Troubles of Theodore Swann* by Edward Griffith and Carolyn Green Satterfield. Griffith worked with Swann during the early days and evidently kept careful records and notes on their interactions, along with correspondence with and about the industrialist. Griffith and Satterfield also used Monsanto Company archives to great advantage to flesh out facts.

Historical contextual background information on cultural events of the decade of the 1920s and 1930s came from Wikipedia.

With regard to Swann's phosphate enterprises and the ecological havoc that he inadvertently wrought, the Environmental Working Group (EWG.org) provided excellent background information on eutrophication and phosphate water pollution. Another interesting Web site maintained by Local 81 of the ICWUC/ UFCW (www.local81c.com) provided information on the history of Swann's Monsanto phosphoric acid.

A word about the Environmental Working Group: It's my firm belief that I could not have written this book without their superb Web archives on PCBs, including the history of Monsanto's PCB endeavors and the health effects of PCBs on humans. In particular, the EWG studies on the PCB burdens of neonates and U.S. adults were invaluable for understanding both the large picture and the facts with which it was colored. Their site was also beautifully designed for intensive data mining, with complete and accessible search functions that I found to be unrivaled save for *Our Stolen Future* by Dr. Theo Colborn and colleagues, Dr. Peter Montague's *Rachel's Democracy & Health News*, the Fox River Watch PCB site, and the online version of the unsurpassed *Environmental Health Perspectives*.

CHAPTER 2 — THE GOOD OL' BOYS OF MONSANTO

For the early history of Monsanto, I consulted the *St. Louis Magazine* Web site, the Mound City on the Mississippi Web site, the Monsanto Historic Archive Collection (online) maintained by Washington University in St. Louis, and Griffith and Satterfield's *The Triumphs and Troubles of Theodore Swann*. Griffith and Satterfield also provided the great majority of the material on the life and times of Theodore Swann, taking the reader to the very moment of the creation of PCBs and on through Swann's tragically fascinating personal and financial destruction.

The "report" of Dr. Frederick Flinn was obtained from legal discovery files archived at the excellent Environmental Working Group Web site, as was early research on chloracne and dermatitis. Data on PCBs and chloracne, all readily available via Internet search engines, came from a number of Web sites. EWG also published electronically the studies by Cecil Drinker completed in the 1930s, the first smoking gun to be found for the lethal effects of PCBs on factory workers. Tangentially, EWG made available the minutes from the seminal conference held on Drinker's studies, including statements by Dr. Emmett Kelly and F. R. Kaimer.

As part of the Nobel e-Museum, the Web site of the Nobel Foundation, there is a nicely written and informative autobiographical piece that touches on Monsanto and PCBs, by Nobel laureate John Fenn, one of the early hires by Ted Swann and a staunch defender of the safety and development of PCBs.

The Queeny family history and the growth of Monsanto, particularly under the aegis of Edgar Queeny, are nicely narrated in *Faith, Hope and $5,000: The Story of Monsanto* by Dan Forrestal.

CHAPTER 3 — THE LONG CON

Nearly all information in this chapter is based on the Environmental Working Group's posting of internal Monsanto memos, correspondence, manuals, and scientific studies in PDF format.

CHAPTER 4 — THE DISCOVERY

Rachel Carson's seminal *Silent Spring*, on the acute ecological dangers presented by DDT, was used for both background and comparison with PCBs. It is a book that must be read by anybody interested in truly understanding how and why this author wrote *Biocidal*. As to her personal and professional history, excellent background data came from Carson's lengthy obituary in the *New York Times*.

Environmental Health Perspectives online provided excellent background data on Carson's work. *EHP* is a peer-reviewed journal sponsored by the National Institute of Environmental Health Sciences. It is also a superb resource that made the writing of *Biocidal* possible. There is more information in *Biocidal* from *EHP* online than from any other source. Their Web site search function is nothing short of phenomenal. The journal papers are richly documented and provided in both HTML and PDF formats in full-text versions—and provided without cost.

My information on Soren Jensen's discovery of PCBs in the environment came from his highly readable and fact-packed account published in *Ambio* in 1972. Of course, his original findings, published as little more than a blurb in the *New Scientist* on December 15, 1966, began the worldwide alarm. For the Monsanto reaction to the Jensen discovery, I relied on the discovery files on the Environmental Working Group Web site.

Material on the Yusho and Yu-Cheng diseases came from papers by Guo and Hsu in *PCBs: Recent Advances in Environmental Toxicology and Health Effects*, edited by Larry W. Robertson and Larry G. Hansen, and from a superior report in the journal *Industrial Health* published in 2003 by the Japanese investigator Takesumi Yoshimura.

The IBT scandal was covered in outstanding depth and with keen insight by veteran investigative journalist Keith Schneider for the *Amicus Journal* in the spring of 1983.

CHAPTER 5 — THE GLOBAL POISON

Nearly all information on Wayland Swain's discovery of the atmospheric transport of PCBs can be found in William Ashworth's superb *The Late, Great Lakes*. Ashworth did his homework, visiting the Great Lakes and conferring with experts on the threats posed to the surrounding ecosystems.

Information on the biotransportation of PCBs came from two sources, essentially, the first being the extraordinary Web site that Theo Colborn and her colleagues developed as a supplement to their seminal book *Our Stolen Future*. The second and, perhaps, even more important resource for explaining the geo- and biodynamics of PCBs is the book itself. Colborn's narrative describing the trip of an imaginary PCB molecule from its manufacture to its resting place in the fat of a polar bear is simply a brilliant piece of expository science. (In a phone interview, Colborn explained that she and her son actually retraced much of the molecule's journey by car across Texas and the Midwest during a summer's research.)

Good data on PCBs found in humans was compiled by the EPA's John S. Stanley in a report published in the *National Human Adipose Tissue Survey* of 1982. (One wonders why such a report has not been duplicated in the intervening years.) The Environmental Working Group also has a series of contemporary studies of PCBs and other industrial toxicants found in human blood serum. UNEP, a United Nations health monitoring organization, produced perhaps the definitive report on PCB distribution in human populations in their *Global Report 2003: Regionally Based Assessment of Persistent Toxic Substances*. As far as documentation that PCBs are not declining, but actually increasing in the environment, Myra Finkelstein's report "Albatross Species Demonstrate Regional Differences in Northern Pacific Marine Contamination" provided highly relevant information on that troubling trend.

CHAPTER 6 — PCBS AND KIDS

Excellent background information on PCB toxicity as it relates to children came from many sources. However two investigators are particularly outstanding in this area. David O. Carpenter of the State University of New York at Albany is the world's foremost toxicologist when it comes to documenting both the subtle and catastrophic characteristics of PCBs. His studies were invaluable. Sheldon Krimsky's *Hormonal Chaos* was a touchstone book, loaded with information on both the scientific and political aspects of PCB contamination in children.

Susan Schantz's studies on PCBs and children were informative. They were especially excellent in providing background on the Asian PCB disasters. As to a direct link between PCBs and testicular cancer, Lennart Hardell's studies, published online in *Environmental Health Perspectives* were nothing short of stunning, if vastly underreported by the scientific and mainstream media. The same can be said of Nils Skakkebaek's work on reproductive deformities in male children. Peter Montague's *Rachel's Democracy & Health News* had good coverage of some of the "insider" aspects of sexual deformities caused by PCBs as well.

On the neurotoxic effects of PCBs on children, of course the gold standard is the work done by the Jacobsons. Lynn Goldman of Johns Hopkins wrote some incisive, wide-angled, editorial-type papers for *Environmental Health Perspectives* online, limning the cultural consequences of PCB contamination of children in the United States. Colleen Moore's book *Silent Scourge* had an unsurpassed chapter on PCBs and their effects on children, along with a keen analysis of other industrial chemical contaminants. Perhaps the most informative and compassionate work on the subject of toxic chemical threats to children used by this author was *In Harm's Way* by Ted Schettler and colleagues, a Web e-book published free of charge by Greater Boston Physicians for Social Responsibility in conjunction with the Clean Water Fund in 2003.

A superlative Internet source was the comprehensive and wonderfully organized Fox River Watch site, the work of citizen activists monitoring PCB cleanup in the Green Bay area of Wisconsin. *EHP* online also had good coverage of PCBs and breast-milk monitoring. The Environmental News Network (ENN) also covered this important issue with well-executed news stories aimed at a lay readership.

CHAPTER 7 — ADULT REALITIES

The most important material in this chapter on the relationship between PCBs and cancer in adults is based on a groundbreaking paper by P. Lichtenstein in the *New England Journal of Medicine* in July 2000. Based on the famous Scandinavian twin studies, Lichtenstein and his interpreters made a robust case for the importance of the environment in cancer etiology. It was a controversial piece of research, with some of the *New York Times'* science writers doing their best to minimize the findings, while the *Washington Post'*s environmental reporters did the opposite. It should be mentioned here that the coverage of health and industrial chemical contamination by the *New York Times* has itself been the subject of controversy. A superb article in the *Nation* (July 6, 1998) by perhaps the best environmental investigative reporter of our times, Mark Dowie, analyzes the surprisingly strong chemical industry slant that key *Times* correspondents continually seem to take on PCBs and other industrial contaminants.

The connection between PCBs and non-Hodgkin's lymphoma is well elucidated in a number of places. Fox River Watch and Our Stolen Future are two Web sites that have plenty of good data. Both are updated regularly. Perhaps the most comprehensive and convincing research on the subject has been done by Dr. Nathaniel Rothman of the National Cancer Institute. He has published in the British medical journal *Lancet* as well as *Environmental Health Perspectives* online. Regarding Parkinson's disease and PCBs, leading scientist Lisa Opanashuk of the University of Rochester is in the vanguard as of this writing.

The Internet offers a huge amount of data on the link between PCBs and immunity problems in adults, with a Google search turning up almost half a million hits. Some of the best work has been done by Swedish investigators and the brilliant, yet down-to-earth Dr. Bruce Blumberg at the University of California at Irvine.

CHAPTER 8 — PCBS, BREAST CANCER, AND HIDDEN AGENDAS

Larry Tye of the *Boston Globe* wrote a very helpful, probing, and sophisticated piece on conflicts of interest at the *New England Journal of Medicine* and also took time to allow me an extensive phone interview on his work regarding Stephen Safe's "conflict" problems. G. Ludewig provided a good analysis of the Ames test and its efficacy for determining the impact on human health of industrial chemical contaminants in Robertson's book *PCBs*.

The groundbreaking and iconoclastic work of Dr. Kirsten Moysich was derived from two phone interviews and a variety of online sources. Her views on media coverage of the PCBs/breast cancer controversy generated by Safe and Hunter are found in "Research Commentary: Thoughts on Recent Finds Regarding Organochlorines and Breast Cancer Risk," *Ribbon* 6, no. 3 (fall 2001). Dr. Yawei Zhang's outstanding work at Yale was available through a paper in the *American Journal of Epidemiology*. I also conducted a phone interview with Zhang.

As usual, the *Our Stolen Future* and Fox River Watch Web sites had excellent data on the subject of research and analysis regarding PCBs and breast cancer.

Regarding the politics of the debate over PCBs and breast cancer, Krimsky's *Hormonal Chaos* provided its usual clear, wide-angled perspective. Dr. Fred vom Saal was kind enough to give me an illuminating interview, as did Dr. Stephen Safe, who was exceptionally accommodating and generous with his time, it must be added. Both of these controversial figures' views on industrial contaminants and their effects on human health are available in online news coverage and scholarly journals. *Environmental Health Perspectives* online was particularly useful because it provides easy access to vom Saal's research. PBS producer and investigative journalist Doug Hamilton, of *Frontline*, did a superior documentary on the subject of industrial chemical contamination titled *Fooling with Nature*. The transcription of his interview with Safe was a critical seam in the fabric of this chapter.

CHAPTER 9 — KILLER WHALES AND THE WEIGHT

Marla Cone's outstanding account of chemical contamination of our hemisphere's northern reaches in *Silent Snow* supplied good data on the problems being encountered by orcas. The folks at Our Stolen Future provided me with Theo Colborn and Michael Smolen's inclusive and informative paper "Epidemiological Analysis of Persistent Organochlorine Contaminants in Cetaceans," which also gave me in-depth information on the dangers that orcas and other whales face from synthetic chemical contamination. But easily the most important data that this chapter is based on is the work of Dr. Peter Ross, the indefatigable and wide-ranging researcher who has almost single-handedly brought to public awareness the dire consequences of PCB pollution for killer whales in the northwest and seals in the Baltic region. Dr. Joseph Cummins can be credited with ringing what was perhaps the original alarm in his writing for the *Ecologist* in 1988. He also allowed this author an extensive interview.

CHAPTER 10 — A LETHAL EROSION OF THE BIOSPHERE

Dr. Arjan Palstra's work with eels and PCBs was perhaps the most elegant and thorough of any regarding the biocidal consequences of PCB contamination. I found his papers on the eel tragedy in the European journal *Naturwissenschaften* not only well written and clear but riveting—and of course more than a bit scary. I also have to thank Dr. Palstra for allowing me to extensively query him about his laboratory research via e-mail.

The *New Scientist*, *National Geographic*, and Our Stolen Future Web site (where I originally discovered Palstra's work) all had informative articles and papers, and, in the case of OSF, excellent links to more information.

Much of my material on lake trout extirpation was obtained in a lengthy and very helpful phone interview with Dr. Deb Swackhamer.

Amphibian decline and its consequences are beautifully and cogently limned by Dr. Robert C. Stebbins and Nathaniel W. Cohen in their seminal *Natural History of Amphibians*.

CHAPTER 11 — THE DEVIL'S GAMBLE

Dr. Barry Commoner's *The Closing Circle* provided superb political and cultural background material for this chapter. He was there and he was active. I found the account of the formation of the EPA tucked away on the Internet in the recollections of family friends via Earth Day founder Denis Hayes in his April 2000 online "Timeline." Dr. Sheldon Krimsky's *Hormonal Chaos* provided invaluable information on the history of the Toxic Substances Control Act (TSCA) and the politics of regulation. The University of Massachusetts-Lowell Center for Sustainable Production (online) had an unsurpassed exposition on the topic as well. The EPA's Web site also contained some relevant information.

The Environmental Defense Fund's 1997 paper *Toxic Ignorance* went into great depth on the failures of governmental regulation of toxic chemicals and was used extensively as source material. Dr. Peter Montague's *Rachel's Democracy & Health News* also did exceptionally fine coverage of the failures of the Clinton administration's EPA and its handling of the Toxic Substances Control Act. Much data in this chapter must also be credited to another superb effort by the Environmental Working Group in its 2005 Web-based critique of TSCA administration by the EPA.

CHAPTER 12 — THE POLITIKS OF PCBS

The Environmental Working Group did what this author can only describe as a tour de force of investigation and data dissemination in their Web site's "Chemical Industry Archives: The Inside Story, Anniston, Alabama." Here they compiled not only the hundreds of discovery documents related to lawsuits filed against Monsanto/Solutia, but they also included investigative journalistic coverage of the Anniston fiasco. The British journal the *Ecologist* published an entire issue on Monsanto's political behavior both in the United States and abroad. I used the articles therein extensively for background in drawing a picture of that corporation's cultural oeuvre. Excerpts from the online version of *The Club* by Sam Smith were employed to document the extent of corporate influence over federal governance.

But by far the most influential source used in this chapter was the redoubtable William Greider's story of the corruptive, corrosive influence of corporations within our government in his book *Who Will Tell the People?* Greider has an uncanny ability to publish tellingly detailed facts and weave them into arguments that go to the very heart of the functioning of contemporary democracy. His is an absolutely indispensable read for any who care to be truly informed citizens about the too often redolent interface of the corporate world and government.

CHAPTER 13 — THE EPIPHANY

The informational foundation of this chapter rests on the book *Our Stolen Future* by Theo Colborn and her colleagues. It may well rank with Carson's *Silent Spring* as germinal to a higher level of human consciousness. And if this book achieves a similar level of acclaim, then Colborn's intellectual synthesis of the endocrine-disrupter theory may well take its place—perhaps a rung or two down from Darwin, Wallace, and Mendel—in the pantheon of great bio-evolutionary breakthroughs.

Along with the book *Our Stolen Future* and the research-packed OSF Web site, Sheldon Krimsky's *Hormonal Chaos* did a superior job of putting the endocrine-disrupter theory into historical, scientific, political, and cultural perspective. His scholarship and organization on the subject were unparalleled, his book another unsung gem of environmental investigative journalism.

As to Dr. Fred vom Saal's low-dose theory, so critical to Colborn's work, I used the database at *Environmental Health Perspectives* online to find a number of finely crafted papers on the subject, two outstanding examples being Wade Welshon's "Large Effects from Small Exposures" and A. Brouwer's "Characterization of Potential Endocrine-Related Health Effects at Low-Dose Levels of Exposure to PCBs." *Rachel's Democracy & Health News* published a wonderfully lucid piece by Colborn coauthor John Peterson Myers, "From Silent Spring to Scientific Revolution." The American Chemistry Council (online) provided data countering that of Colborn and vom Saal in their paper "Chemicals in the Environment and the Endocrine System," as did Dr. Holger Breithaupt in the EMBO report "A Cause without a Disease."

CHAPTER 14—GE AND THE JACKING OF THE HUDSON

The story of the contamination of the Hudson River by PCBs was best related to me by Robert Boyle, the author of *Malignant Neglect*. It was avid Hudson River fisherman Boyle who actually discovered PCBs in the Hudson in the late 1960s as part of an investigative series on marine pollution for *Sports Illustrated*. A renowned river rat, Boyle caught the fish specimens himself from the river and then shipped them off for lab testing. Boyle's fish had massive amounts of polychlorinated biphenyl—one eel Boyle caught had 500 parts per million. In any case, my many phone and e-mail interviews with Bob Boyle were not only edifying, but always fascinating. Into his eighties, Boyle's recall of facts and anecdotes was nothing short of amazing. A wonderful man with a beautiful, springwater-clear mind.

Much of the information on Jack Welch's background comes from *At Any Cost*, Thomas F. Boyle's biography of the General Electric superstar, a marvelously documented and readable book.

Time magazine had excellent coverage of the early battles between activists and GE over PCBs in the Hudson, easily accessible through their voluminous online archives—a terrific boon to understanding the controversy in historical perspective. The *Multinational Monitor* (online) published an informative and well-researched piece by Charlie Cray titled "Toxics on the Hudson" that I used extensively.

Material on Sister Patricia Daly came from a number of online sources, with good examples being the archives of the *Chicago Tribune*, which provided an excellent piece titled "A Nun CEO's Are Recognizing" by Geoff Dougherty, and *Sierra magazine*, whose Marilyn Berlin Snell did a nicely rounded piece on Sister

Pat titled "Sister Action: The Almighty Dollar Meets Its Match in a Dominican Nun."

The Hudson River Sloop Clearwater group, the Hudson River Foundation, the *Poughkeepsie Journal*, Riverkeeper (founded by Bob Boyle), and the New York and D.C. offices of the Natural Resources Defense Council all had usable and informative stories and data online that I used extensively for this chapter and the following chapter.

CHAPTER 15 — THE INEVITABILITY OF NOTHING

Since chapter 15 is a continuation of the previous chapter, many of the same sources were used, along with some exceptional exceptions. Amanda Griscom Little's "GE's Green Gamble" and Richard Pollak's "Is GE Mightier Than the Hudson?" were extensively used in my research in this follow-up chapter. Wikipedia also provided good background information on some of the players, such as Christine Whitman and Ann Gorsuch. The Associated Press provided excellent coverage of GE shareholder meetings and Sister Daly's forays. In that regard, *Harper's* magazine published an invaluable article, "God Versus GE," containing actual transcripts of Sister Pat's shareholder meeting jousts with Jack Welch. Daly's own press release on GE's expenditures was also highly illuminating. Dan Shapley of the *Poughkeepsie Journal* wrote consistently excellent pieces on human illness and ecological damage associated with PCB contamination that I referred to in this chapter, as did the staff writer/researchers for the Hudson River Foundation: Joel E. Baker, W. Frank Bohlen, Richard Bopp, Bruce Brownawell, Tracy K. Collier, Kevin J. Farley, and W. Rockwell Geyer. Salon's David Neiwert did good "insider" Washington work with his piece on Ted Olson, as did Daren Fonda, for *Time* magazine, in his fine piece "GE's Green Awakening"; and I used Charlie Cray's superb investigative reporting for the *Multinational Monitor* ("Toxics on the Hudson," summer of 2001) for this chapter as well as chapter 14.

CHAPTER 16 — PRECAUTIONARY AGONISTES

This chapter evolved largely from superb articles in the *Multinational Monitor* and on the Science and Environmental Health Network Web site by Nancy Meyers in September of 2004. Equally informative and incisive was Peter Montague's writing in his *Rachel's Precaution Reporter*. Montague has the unique ability to consistently synthesize simple, strong arguments from highly complex data and theory. Mary O'Brien's paper "Contemplating Impulse and Acting on Our Navels" was the best wide-angled, yet detailed overview (along with references) of the precautionary principle that I was able to find online. In a lengthy phone interview, I wrangled her into sending me her speech "Beyond Democratization of Risk Assessment." It provided a terrific view of how an insider looks at the practical problems facing the real-world use of risk assessment.

Carl Cranor's book *Regulating Toxic Substances* was certainly helpful.

Frank Ackerman and Rachel Massey, in their *Prospering with Precaution*, made an excellent case for the precautionary principle as a compelling economic policy. Anti–junk-food guru Michael Pollan had a well-done and streamlined editorial in the *New York Times* that went directly to the crux of the complicated subject of the politics of the precautionary principle but was still highly readable, as is all his fine writing.

CHAPTER 17 — EPIGENETICS, PCBS, AND US

The best background material for this chapter came from the PBS/WGBH Web site that supplemented the *Nova* broadcast "The Ghost in Your Genes." Alas, nowhere on the Web site, at least that I could find, were there any producer credits, so unfortunately I can't credit anyone for this brilliant piece of journalism. However, as of this writing, the PBS Web site for the documentary is still going strong with transcripts, interviews, and data available for download. The Human Epigenome Project (online) had good general information and provided a fine overview and simplified explanations of the dynamics and impact of epigenetic research.

As far as facts and research, by far the most important resource for this author was the Web site of Matthew Anway and Michael Skinner of the Center for Reproductive Biology at the School of Molecular Biosciences, Washington State University. Their work spearheaded the birth of epigenetics as a revolutionary force in the biological sciences. Dr. Anway was also very kind to allow me a lengthy interview on his work given the pressures of his schedule.

Dr. David Crews of the University of Texas at Austin was also helpful in his explanations of just how PCBs fit into the epigenetic picture via phone interviews. Michael Balter did a nice "Daily News" piece on the *Science* magazine Web site on the importance of Crews and his colleague Andrea Gore's experiments confirming the findings of Anway and Skinner. Perhaps the most illuminating popular writing on the subject, for this author at least, was Jane Bradbury's "The Human Epigenome Project—Up and Running," published on the *PLoS Biology* site, with excellent links and resources provided.

EPILOGUE: CLOUDS AND SUNLIGHT

With one exception, all data for this chapter comes from previously mentioned sources and personal interviews done for this chapter specifically, such as those with Medina Electric's Mark Rollans. The only new source used for this chapter was Ray Kurzweil's *The Singularity Is Near*. However, Grandjean and Landrigan's paper for the *Lancet* was truly the centerpiece of this chapter. For all those interested in understanding the current and future threat posed by industrial chemicals, go online and spend thirty dollars or so to download this article. It is probably the most important document written on the subject since Ted Swann had the first batch of the "magic fluid" manufactured.

Bibliography

Ahmad, A. "Killer Whales Most Contaminated by PCBs." Coalition Opposed to PCB Ash (COPA) online, February 25, 2003.

Alava, J., et al. "Loggerhead Sea Turtle (Caretta caretta) Egg Yolk Concentrations of Persistent Organic Pollutants." *Science of the Total Environment* 367 (2006).

Allen, A. "The Autism Numbers: Why There's No Epidemic." Slate.com, January 15, 2007.

Ambrosone, C., et al. "CYP17 Genetic Polymorphism, Breast Cancer, and Breast Cancer Risk Factors." *Breast Cancer Research* 5, no. 2 (2003).

American Cancer Society. "Detailed Guide: Lymphoma, Non-Hodgkin Type." www.cancer.org.

American Chemistry Council. "Comments on the National Toxicology Program Endocrine Disruptors Low-Dose Peer Review Final Report." www.american chemistry.com. July 16, 2001.

———. "American Chemistry Council Perspective on the Low-Dose Hypothesis." www.americanchemistry.com.

———. "A Summary of the Endocrine Disruption Hypothesis."

Ames, Bruce N. "The Causes and Prevention of Cancer." http://users.rcn.com/ jkimball.ma.ultranet/BiologyPages/A/Ames_Causes.html. March 15, 1997.

Anway, Matthew D., et al. "Epigenetic Transgenerational Actions of Endocrine Disruptors and Male Fertility." *Science* 308 (2005).

———. "Endocrine Disruptor Vinclozolin Induced Epigenetic Transgenerational Adult-Onset Disease." *Endocrinology* 147, no. 12 (2006).

Arctic Monitoring and Assessment Programme. "Persistent Organic Pollutants in the Arctic." *AMAP Assessment 2002*, 2004.

Aschwanden, C. "Is Salmon Good for You?" *Natural Solutions*, June 1, 2005. www .naturalsolutionsmag.com/articles-display/8371/Is-Salmon-Good-For-You.

Ashworth, W. *The Late, Great Lakes*. Toronto: Collins, 1986.

Ashworth, W., and C. E. Little. "Mechanics of Toxic Precipitation," in *Encyclopedia of Environmental Studies*, revised edition. New York: Facts on File, 2001.

Associated Press. "EPA Supports Controversial Hudson River Dredging." December 6, 2000.

———. "GE Consultant Questions Effects." April 15, 2001.

———. "Studies Reaffirm PCBs Hazards." April 15, 2001.

———. "EPA Now Says Dredging Won't Start Until 2008." July 28, 2006.

Auman, H. J., et al. "Plastic Ingestion by Laysan Albatross Chicks on Sand Island, Midway Atoll, in 1994 and 1995." In *Albatross Biology and Conservation*. Chipping Norton, UK: Surrey, Beatty and Sons, 1997.

Bakalar, N. "Rise in Rate of Twin Births May Be Tied to Dairy Case." *New York Times*, May 30, 2006.

———. "Testicular Cancer Success Has Doctors Asking Why." *New York Times*, August 1, 2006.

Ballweg, M. "Endometriosis: A New Picture of the Disease Is Emerging." In *The Endometriosis Sourcebook*. Chicago: Contemporary Books, 1995.

"Barge Carrying PCBs Sailed Through Legal Loophole." Canadian Broadcasting Corp. News online, November 27, 2006.

Balter, M. "Toxic Hand-Me-Down." *Science*NOW Daily News, March 27, 2007.

Barlow, J. "Heavy Consumption of Tainted Fish Curbs Adult Learning and Memory." University of Illinois News Bureau, June 1, 2001.

Barrett, D. "U.S. Asks Judge to Approve Hudson Dredging Agreement." Associated Press, May 16, 2006.

Barringer, F. "Polluted Sites Could Face Shortage of Cleanup Money." *New York Times*, August, 15, 2004.

———. "A Coalition for Firm Limit on Emissions." *New York Times*, January 19, 2007.

Batt, S., and L. Gross. "Cancer, Inc." *Sierra*, September/October 1999.

BBC—Health online. "Tainted Fish Harm Anglers' Brains." June 6, 2001.

———. "Pollutants Affect Babies' Brains." November 9, 2001.

———. "Pollutants May Produce More Boys." January 29, 2002.

———. "Pollution 'Adds to Sun Cancer Risk." March 22, 2002.

———. "Toxic Chemicals That Poisoned Your Great-Grandparents May Also Damage Your Health, U.S. Research Suggests." June 3, 2005.

Beatty, J. "The Resignation Principle: An Open Letter to Christie Whitman." *Atlantic Monthly*, July 17, 2003.

Beck, L. "Food for Thought." (Toronto) *Globe and Mail*, October 18, 2006.

Bedard, D. L. "Microbial Dechlorination of PCBs in Aquatic Sediments." In *PCBs: Recent Advances in Environmental Toxicology and Health Effects*, eds. Robertson and Hansen.

Beland, P., et al. "Toxic Compounds and Health and Reproductive Effects in St. Lawrence Beluga Whales." *Journal of Great Lakes Research* 19, no. 4 (1993).

Bergeron, J. M., et al. "PCBs as Environmental Estrogens." *Environmental Health Perspectives* 102, no. 9 (September 1994).

Betts, K. "Endocrine Disrupters Ubiquitous in U.S. Homes." *Environmental Science & Technology* 37, no. 21 (September 2003).

Bhalla, N. "Whales Reveal Man's Damaging Impact on Oceans." Reuters, October 12, 2003.

Biever, C. "Autism: Lots of Clues, But Still No Answers." *New Scientist,* May 14, 2005.

Birnbaum, L. F. E. "Cancer and Developmental Exposure to Endocrine Disruptors." *Environmental Health Perspectives* 111, no. 4 (2003).

Black, Harvey. "Vitamin E vs. PCBs." *Environmental Health Perspectives* 108, no. 1 (January 2000).

Blaustein, A., et al. "Ultraviolet Radiation, Toxic Chemicals and Amphibian Population Declines." *Diversity and Distributions* 9, no. 2 (March 2003).

Bluemink, E. "Researcher at Trial: PCBs Can Increase Disease Risk." *Anniston* (Alabama) *Star,* January 23, 2002.

———. "More Problems for Solutia, the EPA? Public Comment Surrounding. PCB Case under Federal Review." *Anniston Star,* June 17, 2002.

Blumberg, B., et al. (2004). "Highly Chlorinated PCBs Inhibit the Human Xenobiotic Response Mediated by the Steroid and Xenobiotic Receptor (SXR)." *Environmental Health Perspectives* 112, no. 2 (February 2004).

Borowski, J. F. "Christine Todd Whitman: Why Are You Still a Lapdog?" CommonDreams.org, June 30, 2003.

Bowen, K. "The PCB Legacy: Question of PCB Risks Uncertain." *Rome* (Georgia) *News-Tribune,* September 9, 2003.

Bowman, L. "Benefits Outweigh Risks, So Eat More Fish." Scripps Howard News Service, October 18, 2006.

Boyle, R. H. "Hudson River Lives," *Audubon* 73, no. 2 (March 1971).

Boyle, R. H., and Environmental Defense Fund. *Malignant Neglect.* New York: Knopf, 1979.

Breithaupt, H. "A Cause without a Disease." *EMBO* [European Molecular Biology Organization] *Reports* 5 (1 doi:10.1038/sj.embor.7400063).

Brenner, B. "Media Watch: A Safe View of DDT." Breast Cancer Action online newsletter, no. 45 (December/January 1997/98).

Brink, V., et al. "Multivariate Analysis of Ecotoxicological Data Using Ordination: Demonstrations of Utility on the Basis of Various Examples." *Australasian Journal of Ecotoxicology* 9 (2003).

Brouwer, A. "Report of the WHO Working Group on the Assessment of Health Risks for Human Infants from Exposure to PCDDs, PCDFs and PCBs." *Chemosphere* 37, nos. 9–12 (October–November 1998).

Brouwer, A., et al. "Characterization of Potential Endocrine-Related Health Effects at Low-Dose Levels of Exposure to PCBs." *Environmental Health Perspectives* 107, suppl. 4 (August 1999).

Brown, S., et al. "Altered Thyroid Status in Lake Trout (Salvelinus namaycush) Exposed to Co-Planar 3,3,'4,4,'5-pentachlorobiphenyl" (abstract). *Aquatic Toxicology* 67, no. 1 (March 30, 2004).

Brown, T. M., et al. "Reduction of PCB Contamination in Arctic Coastal Environment." *Environmental Science and Technology* 43, no. 20 (September 15, 2009).

Browner, C. M. "Oral Statement before the Committee on Environmental Conservation New York State Assembly." EPA news release, July 9, 1998.

Brucker-Davis, F. "Effects of Environmental Synthetic Chemicals on Thyroid Function." *Thyroid* 8 (1998).

Brundell, R. "Veolia Transporting Toxic Waste to Europe." *Business News Americas*, www.bnamericas.com, September 23, 2005.

Bruno, K. "Monsanto's Failing PR Strategy." *Ecologist* 28, no. 5 (1998).

Bueckert, Dennis. "New Tests Find Poison in Children's Blood, Urine." *Canadian Press.* http://list.web.net/archives/sludgewatch-l/2006-June.txt, June 1, 2006.

Burnstein, H. "Roots of Cancer: Nature or Nurture." *New England Journal of Medicine* 343, no. 2 (2000).

Burros, M. "Farmed Salmon Is Said to Contain High PCB Levels." *New York Times*, July 30, 2003.

———. "Is There an Extra Ingredient in Nonstick Pans?" *New York Times*, July 27, 2005.

———. "One Study Calls Fish a Lifesaver, Another Is More Cautious." *New York Times*, October 18, 2006.

———. "US: Advisories on Fish and the Pitfalls of Good Intent." *New York Times*, February 15, 2006.

Burton, B. "PCB Knowledge Flows from Indoor Stream." Environmental News Service, April 10, 1998.

Busbee, D. "PCBs in Texas Dolphins a Warning to Humans." Environmental News Service, February 19, 1999.

Butterworth, T. "How Activist Groups Run the News." STATS.org (Statistical Assessment Service), http://stats.org, January 31, 2007.

Calamai, P. "Dead Salmon Boost Toxins in Lakes." *Toronto Star*, September 18, 2003.

Calle, E., et al. "Organochlorines and Breast Cancer Risk." *CA: A Cancer Journal for Clinicians* 52, no. 301 (2002).

"Cancer: Nature, Nurture, or Both" (editorial). *New England Journal of Medicine* 343, no. 2 (2000).

Cantor, K., et al. "Risk of Non-Hodgkin's Lymphoma and Prediagnostic Serum

Organochlorines." *Encyclopedia Health Perspectives online* 111, no. 2 (February 2003).

Carey, B. "Study Puts Rate of Autism at 1 in 150 U.S. Children." *New York Times*, February 9, 2007.

Carey, C., et al. "Amphibian Declines and Environmental Change: Use of Remote-Sensing Data to Identify Environmental Correlates." *Conservation Biology* 15, no. 4 (August 2001).

Carlowicz, M. "Mistaken Identity: Two Bromine Compounds Found in Whale Blubber Are Natural Products, Not Industrial Pollutants." *Oceanus*, February 10, 2005. www.whoi.edu.

"Carol M. Browner: Biography." EPA Web site, February 1999.

Carpenter, D. O. "Polychlorinated Biphenyls and Human Health." *International Journal of Occupational Medicine and Environmental Health* 11 (November 4, 1998).

———. "Studies Link PCBs to Human Cancer." *Wall Street Journal*, January 4, 2001.

Carpenter, D. O., et al. "Understanding the Human Health Effects of Chemical Mixtures." *Environmental Health Perspectives* 110, suppl. 1 (February 2002).

Carpenter, M. "Amphibian Fungus Found on Frogs in Maine." *New York Times*, July 4, 2006.

Carson, R. *Silent Spring.* New York: Houghton Mifflin, 1962.

CAS [Chemical Abstract Service] Registry. "Aroclor 1254." National Institute of Environmental Health Sciences, 2006.

Casten, L. C. "An ECO-LOGICal Way to Dispose of Waste." *Environmental Health Perspectives* 103, no. 9 (September 1995).

"Caulking Found to Be Unrecognized Source of PCB Contamination in Schools and Other Buildings." *Science Daily*, July 26, 2004.

Center for Health Effects of Environmental Contamination. "Endocrine Disruptors and Pharmaceutical Active Compounds in Drinking Water Workshop," report of proceedings, University of Iowa, April 19–21, 2000.

Centers for Disease Control. "Second National Report on Human Exposure to Environmental Chemicals." www.cdc.gov/exposurereport, January 2003.

———. "Are DES Sons at an Increased Risk for Cancer?" www.cdc.gov/DES/consumers/about/concerns_sons.html, 2006.

Cernd, M., et al. "PCB Congeners, PCDDs, and PCDFs in the Adipose Tissue of the Czech Population." In *PCBs: Recent Advances in Environmental Toxicology and Health Effects*, eds. Robertson and Hansen.

Chang, H., et al. "Transgenerational Epigenetic Imprinting of the Male Germline by Endocrine Disruptor Exposure during Gonadal Sex Determination." *Endocrinology* 147, no. 12 (2006).

"Chemicals Harm Children before They Are Born" (editorial). *ToxCat* 2, no. 3 (summer 1996).

Children's Environmental Health Network. "Endocrine Disrupters Position Statement." Children's Environmental Health Network, 2006.

Chlorine Chemistry Division. "Response to Our Stolen Future." www.american chemistry.com/chlorine, 1996.

———. "What Are the Facts about Dioxins and Furans?" www.americanchemistry .com/chlorine, 2006.

Clean Water Action Council. "Women Who Eat Polluted Fish Increase Their Breast Cancer Risks." News release, November 7, 2003.

———. "Clues from Wildlife to Create an Assay for Thyroid System Disruption." *Environmental Health Perspectives* 110, suppl. 3 (June 2002).

———. "Superfund Money Crisis Slows Fox River Cleanup."

Cogliano, V. J. (2001). "Considerations for Setting Reference Values for Environmental PCBs." In *PCBs: Recent Advances in Environmental Toxicology and Health Effects*, eds. Robertson and Hansen.

Cohen, B. "Big Government and Bad Science." Institute for Policy Innovation Publication. 1999.

———. "The Environmental Working Group: Peddlers of Fear." www.ewg.org, 2004.

Cohen, B. R. "EPA Will Destroy Hudson River to Save It." *Wall Street Journal*, December 21, 2001.

Cohen, J. "Studies Find Heart Deformities, Higher Mortality Rates in PCB-Affected Wildlife." Indiana University Media Relations, April 3, 2006.

Coifman, J. "With Important Decisions Nearing, Officials Withhold over a Thousand Documents." Natural Resources Defense Council, news release, April 6, 2006.

Colborn, T. *Great Lakes: Great Legacy?* Washington, DC: The Conservation Foundation. 1990.

———. "The Wildlife/Human Connection: Modernizing Risk Decisions." *Environmental Health Perspectives* 102, suppl. 12 (December 1994).

———. "Epidemiological Analysis of Persistent Organochlorine Contaminants in Cetaceans." *Reviews of Environmental Contamination and Toxicology* 146 (1996).

Colborn, T., et al. (1993). "Developmental Effects of Endocrine-Disrupting Chemicals in Wildlife and Humans." *Environmental Health Perspectives* 101, no. 5 (October 1, 1993).

Colborn, T., D. Dumanoski, and J. P. Myers. *Our Stolen Future*. New York: Plume Books, 1997.

Collins, J., et al. "Global Amphibian Declines: Sorting the Hypotheses." *Diversity and Distributions* 9 (2003).

Commoner, B. *The Closing Circle*. New York: Bantam Books, 1974.

Cone, M. "A Disturbing Whale Watch in Northwest." *Los Angeles Times*, February 16, 2001.

————. *Silent Snow*. New York: Grove Press, 2005.

"Contaminated Puget Sound Food Chain Blamed for Whale Decline." Environmental News Service, December 17, 2001.

Cook, P., et al. "Effects of Aryl Hydrocarbon Receptor-Mediated Early Life Stage Toxicity on Lake Trout Populations in Lake Ontario during the 20th Century." *Environmental Science and Technology* 37 (2003).

Cooke, P. S. "Disruption of Steroid Hormone Signaling by PCB." In *PCBs: Recent Advances in Environmental Toxicology and Health Effects*, eds. Robertson and Hansen.

Coombs, A. "Pollutant Levels Rising in Open Water." *Science Now*, April 4, 2006.

Cope, G. "America's #1 Superfund Polluter: General Electric." U.S. PIRG, www .pirg.org, 2003.

CORDIS (Community Research and Development Information Service—Europe). "Postnatal Exposure to PCBs Found to Depress Immune Responses." August 22, 2006.

Corn, D. "Whitman Warms the Bench." *Mother Jones*, March 20, 2002.

CorpWatch. "General Electric." www.corpwatch.org.

Cortese, A. "Dupont, Now in the Frying Pan." *New York Times*, August 8, 2004.

Cranor, C. *Regulating Toxic Substances*. New York: Oxford University Press, 1993.

Cray, C. "Toxics on the Hudson: The Saga of GE, PCBs and the Hudson River." *Multinational Monitor* 22, no. 7 (July/August 2001).

Crews, D. "Epigenetics and Its Implications for Behavioral Neuroendocrinology." Special issue, *Epigenetics* 29, no. 3, "Frontiers in Neuroendocrinology" (2008).

Crews, D., et al. "Endocrine Disruptors: Present Issues, Future Directions." *Quarterly Review of Biology* 75, no. 3 (2000).

————. "Epigenetics, Evolution, Endocrine Disruption, Health, and Disease." *Endocrinology* 147, no. 6 (2006).

————. "Transgenerational Epigenetic Imprints on Mate Preference." *PNAS (Proceedings of the National Academy of Sciences)* 104, no. 14 (April 3, 2007).

Crowe, K. "Dredging 'Tour' Clarifies EPA Plan." *Times Union* (Albany, NY), June 16, 2006.

Cummins, J. E. "Extinction: The PCB Threat to Marine Mammals." *Ecologist* 18 (November 1988).

————. "PCBs: Can The World's Sea Mammals Survive Them?" *Ecologist* 28, no. 5 (1998).

Cushman, J. H. "Study Finds Little Risk from PCBs." *New York Times*, March 10, 1999.

Daszak, P., et al. "Anthropogenic Environmental Change and the Emergence of Infectious Diseases in Wildlife." *Acta Tropica* 78 (2001).

Datz, T. "New Study Shows the Benefits of Eating Fish Greatly Outweigh the Risks." Harvard School of Public Health, news release, October 17, 2006.

Davidson, O. G. *Fire in the Turtle House*. New York: Public Affairs, 2001.

Demers, A., et al. "Plasma Concentrations of PCBs and the Risk of Breast Cancer: A Congener-specific Analysis." *American Journal of Epidemiology* 155, no. 7 (2002).

Denham, M., et al. "Relationship of Lead, Mercury, Mirex, Dichlorodiphenyldichloroethylene, Hexachlorobenzene, and Polychlorinated Biphenyls to Timing of Menarche among Akwesasne Mohawk Girls." *Pediatrics 115-2* (February 2005).

Denver, R., et al. "Effects of PCBs on Great Lakes Amphibians." Michigan Department of Environmental Quality, 2005.

DePalma, A. "GE Commits to Dredging 43 Miles of Hudson River." *New York Times*, October 7, 2005.

DeRoos, A., et al. "Persistent Organochlorine Chemicals in Plasma and Risk of Non-Hodgkin's Lymphoma." *Cancer Research* 65 (December 1, 2005).

DeSolla, S., et al. "Impact of Organochlorine Contamination on Levels of Sex Hormones and External Morphology of Common Snapping Turtles." *Environmental Health Perspectives* 106, no. 5 (May 1998).

DeSwart, R., et al. "Impaired Immunity in Harbour Seals (Phoca vitulina) Exposed to Bioaccumulated Environmental Contaminants: Review of a Long-term Feeding Study." *Environmental Health Perspectives* 104, suppl. 4 (August 1996).

Dew, J. "A link between PCB Pollution in the Berkshires and Developmental Disabilities" [*Berkshire* (Mass.) *Eagle*, March 18, 2004.

———. "Two Studies Document PCB Harm." *Berkshire Eagle*, June 20, 2005.

———. "State to Test PCBs in Blood." *Berkshire Eagle*, May 4, 2006.

"Dioxin and PCBs in Four Commercially Important Pelagic Fish Stocks in the North East Atlantic." *Nordisk Atlantsamarbejde*, April 2003.

Dold, C. "Toxic Agents Found to Be Killing Whales." *New Sunday Times* (Malaysia), June 21, 1992.

Dowie, M. "Stormy Weather." Media Channel.org, February 17, 2000.

Doyle, A. "Norway Advises Pregnant Women against Whale Meat." Reuters, May 12, 2003.

"Dredging Quandary" (editorial). *Times Union* (Albany, NY), July 20, 2006.

Drinker, C. "The Problem of Possible Systemic Effects from Certain Chlorinated Hydrocarbons." *Journal of Industrial Hygiene and Toxicology* 19, no. 7 (1937).

Drukier, C. "Abnormal Birth Rates in Canadian Native Reserve." *Epoch Times* (Canada), June 9, 2006.

Dubose, L. "Whacked by Whitman." *Texas Observer*, May 24, 2002.

Dujardin, M., et al. "The Belgian "Dioxin" Crisis." In *PCBs: Recent Advances in Environmental Toxicology and Health Effects*, eds. Robertson and Hansen.

Dunham, W. "When Pregnant Mom Eats Fish, Kids Do Better: Study." *Science News*, February 15, 2007.

Durodie, B. "Gender-bending Chemicals: Facts and Fiction." Spiked Science online, July 11, 2001.

Easterbrook, G. "Environmental Doomsday: Bad News Good, Good News Bad." *Brookings Review* 20, no. 2 (2002).

Efron, E. *The Apocalyptics*. New York: Simon & Schuster, 1984.

Eisler, R. "Polychlorinated Biphenyl Hazards to Fish, Wildlife, and Invertebrates: A Synoptic Review." Biological Report no. 85 (1.7), Contaminant Hazard Reviews. Washington, D.C.: U.S. Fish and Wildlife Service, April 1986.

Ekbom, A. "Age at Immigration and Duration of Stay in Relation to Risk for Testicular Cancer among Finnish Immigrants in Sweden." *Journal of the National Cancer Institute* 95 (2003).

El Amin, A. "Dioxins, PCBs, Metals Still at "Safe" Levels, UK Regulator Says." FoodNavigator.com, July 27, 2006.

Elskus, A. A. "Toxicant Resistance in Wildlife: Fish Populations." In *PCBs: Recent Advances in Environmental Toxicology and Health Effects*, eds. Robertson and Hansen.

Environmental Defense Fund. "Dioxins in Fish and Shellfish." www.edf.org; posted: August 12, 2004; updated: September 7, 2004.

Environmental News Service. "General Electric Offers Hudson River Settlement." April 11, 2002.

———. "EPA Ombudsman Resigns." April 23, 2002.

"EPA Orders Dredging of New York's Hudson River." CNN.com, December 5, 2001.

"Epigenetic Mechanisms and Gene Networks in the Nervous System" (symposium). *Journal of Neuroscience* 25, no. 45 (November 9, 2005).

Epstein, S. "Unlabeled Milk from Cows Treated with Biosynthetic Growth Hormones." *International Journal of Health Services* 26, no. 1 (1996).

Erickson, M. D. "PCB Properties, Uses, Occurrence, and Regulatory History." In *PCBs: Recent Advances in Environmental Toxicology and Health Effects*, eds. Robertson and Hansen.

"Everything You Never Wanted to Know about Monsanto's M.O." Mindfully .org, 2006.

Ewald, G., et al. "Biotransport of Organic Pollutants to an Inland Alaska Lake by Migrating Sockeye Salmon." *Arctic* 51 (1998).

Fagin, D. *Toxic Deception*. New York: Birch Lane Press, 1996.

Faroon, O., et al. "Polychlorinated Biphenyls: Human Health Aspects." WHO (World Health Organization) online, 2003.

Feinstein, H. "Delivery of Pollutants by Spawning Salmon." *Nature* 425, no. 18 (September 2003).

Feist, G., et al. "Evidence of Detrimental Effects of Environmental Contaminants on Growth and Reproductive Physiology of White Sturgeon in Impounded Areas of the Columbia River." *Environmental Health Perspectives* 113, no. 12 (December 2005).

Fenn, John B. "Autobiography." Nobel Prize Web site, nobelprize.org/nobel_prizes/chemistry/laureates/2002/fenn-autobio.html.

Ferrara, J. "Revolving Doors: Monsanto and the Regulators." *Ecologist* 28, no. 5 (1998).

Feynman, R. "Cargo Cult Science." California Institute of Technology commencement address, 1974.

Fiedler, H. "Safe Management of PCB and Case Studies." UNEP (United Nations Environment Program), www.chem.unep.ch/Pops/POPs_Inc/proceedings/stpetbrg/fidler2.htm.

———. "Global and Local Disposition of PCBs." In *PCBs: Recent Advances in Environmental Toxicology and Health Effects*, eds. Robertson and Hansen.

Fields, S. "Great Lakes: Resource at Risk." *Environmental Health Perspectives* 113, no. 3 (March 2005).

Finkelstein, M. "Albatross Study Shows Regional Differences in Ocean Contamination." *University of California at Santa Cruz Environmental News Letter*, April 4, 2006.

Finkelstein, M., et al. "Albatross Species Demonstrate Regional Differences in North Pacific Marine Contamination." *Ecological Applications* 16, no. 2 (2006).

Finney, K. "Autism Conference Reports Advances." University of California at Davis M.I.N.D. Institute, news release, May 5, 2005.

Fisch, H., et al. "The Possible Effects of Environmental Estrogen Disrupters on Reproductive Health." *Current Urology Reports* 1 (2000).

Fisher, J. "Environmental Anti-Androgens and Male Reproductive Health: Focus on Phthalates and Testicular Dysgenesis Syndrome." *Reproduction* 127 (2004).

Fitz-Gibbon, J. "General Electric Hires First Two Contractors for Huge HudsonRiver PCBs Dredging Project." *Journal News* (Lower HudsonValley, NY), January 27, 2007.

Flanders, L. "Don't Cry for Christie." *San Francisco Chronicle*, May 22, 2003.

Fleming, P. "EPA Releases Human Health Risk Assessment for GE Pittsfield/Housatonic River Site." Environmental Protection Agency, news release, April 3, 2006.

"Food for Thought." *Science News*, September 14, 1996.

Forman, D., et al. "Aetiology of Testicular Cancer: Association with Congenital Abnormalities, Age at Puberty, Infertility, and Exercise." *British Medical Journal* 308 (May 28, 1994).

Forrestal, D. J. *Faith, Hope and $5,000: The Story of Monsanto.* New York: Simon & Schuster, 1977.

Fournier, M., et al. "Validation of an Amphibian Model to Assess the Effects of Persistent Organic Pollutants on Amphibian Physiology." *Health Canada,* June 23, 2004.

Fox, G. "Wildlife as Sentinels of Human Health Effects in the Great Lakes–St. Lawrence Basin." *Environmental Health Perspectives* 109, suppl. 6 (December 2001).

Fox, M. "Men Carrying Pollutants Have More Boys: U.S. Study." Reuters, January 29, 2002.

———. "Dead Worms Show River Cleanup Works." Reuters, May 5, 2005.

Fox River Watch. "The History of PCBs" (parts 1 and 2). Foxriverwatch.org, 2006.

———. "PCB Baby Studies." (part 2). Foxriverwatch.org.

Frame, G. M. "The Current State of the Art." In *PCBs: Recent Advances in Environmental Toxicology and Health Effects,* eds. Robertson and Hansen.

Francis, E. (1994). "Conspiracy of Silence." *Sierra,* September/October 1994.

———. "Buy Me a River." Planet Waves (blog), October 2000.

———. "No Ordinary Lie." Planet Waves, 2003.

Fritsche, E., et al. "Polychlorinated Biphenyls Disturb Differentiation of Normal Human Neural Progenitor Cells: Clue for Involvement of Thyroid Hormone Receptors." *Environmental Health Perspectives* 113(7) (July 7, 2005).

Froescheis, O., et al. "The Deep-Sea as a Final Global Sink of Semivolatile Persistent Organic Pollutants?" *Chemosphere* 40, no. 6 (2000).

Fry, M. "PCBs Still Affecting Bird Populations." Environmental News Service, December 17, 1997.

Fumento, M. "Superfund: Hazardous Waste? It Has Spent Billions but Cleaned Just 63 Sites." *Investor's Business Daily,* October 22, 1991.

Fund, J. O. "Country Report for Japan." PCB symposium, Malaysia, Japan Offspring Fund, 2003.

Garcia, M. "Leaking Lights in Schools Could Pose Health Risks." *Lahontan* (Nevada) *Valley News,* June 24, 2006, www.lahontanvalleynews.com.

Gardner, T. "Declining Amphibian Populations: A Global Phenomenon in Conservation Biology." *Animal Biodiversity and Conservation* 24, no. 2 (2001).

———. "Hot Trash-to-Fuel Technology Gathering Steam." Reuters, February 27, 2004.

Gauger, K., et al. "Polychlorinated Biphenyls (PCBs) Exert Thyroid Hormone-like Effects in the Fetal Rat Brain but Do Not Bind to Thyroid Hormone Receptors." *Environmental Health Perspectives* 112, no. 5 (April 2004).

Gaynor, K. A. "News and Analysis: Environmental Enforcement Developments in 2003." *Environmental Law Reporter* 34 (2003).

Gellerman, B. "Whitman Looks Back." *Living on Earth*, radio broadcast, February 11, 2005.

Giesy, J., et al. "Dioxin-Like and Non-Dioxin-Like Toxic Effects of Polychlorinated Biphenyls (PCBs): Implications for Risk Assessment." *Critical Reviews in Toxicology* 28, no. 6 (1998).

Ginneken, V., et al."Presence of Eel Viruses in Eel Species from Various Geographic Regions." *Bulletin of the European Association of Fish Pathologists 24*, no. 5 (2004).

——. "Eel Migration to the Sargasso: Remarkably High Swimming Efficiency and Low Energy Costs." *Journal of Experimental Biology* 208 (2005).

——. "Gonad Development and Spawning Behaviour of Artificially Matured European Eel." *Animal Biology* 55, no. 3 (2005).

——. "Hematology Patterns of Migrating European Eels and the Role of Evex Virus." *Comparative Biochemistry and Physiology*, 140, part C (2005).

——. "Microelectronic Detection of Activity Level and Magnetic Orientation of Yellow European Eel, Anguilla anguilla l., in a Pond." *Environmental Biology of Fishes* 72 (2005).

Gladden, B., et al. "Prenatal Exposure to DDE and PCB May Affect Body Size at Puberty—Boys Are Taller, Heavier, and Some Girls Are Heavier." *Pediatrics* 136 (April 2000).

Global Program of Action. "How POPs Accumulate in Fish." GPA Clearing House, www.gpa.unep.org, July 11, 2005.

Glynn, A., et al. "Organochlorines in Swedish Women: Determinants of Serum Concentrations." *Environment and Health Perspectives* 111, no. 3 (March 2003).

Godtfredsen, K., et al. "Immunocompetence of Juvenile Chinook Salmon Following Exposure to Dietary PCBs: Implications for Regulatory Policy." *Puget Sound Research*, www.portal.windwardenv.com, 2001.

"God versus G.E." *Harper's*, August 1998.

Gold, C. "Hormone Hell." *Discover* 19, no. 9 (1996).

Goldschmidt, T. *Darwin's Dream Pond*. Cambridge, MA: MIT Press, 1996.

Goldsmith, Z. "Eco Warriors or Vandals? Who Are the Real Terrorists?" *Ecologist* 28, no. 5 (1998).

Goldstein, R. "Riverkeeper Comments on EPA-GE Consent Decree." Riverkeeper.org, December 14, 2006.

Gomez, D., et al. "Number of Boys Born to Men Exposed to Polychlorinated Biphenyls." *Lancet* 360 (2002).

Goodman, B. "To Stem Widespread Extinction, Scientists Airlift Frogs in Carry-on Bags." *New York Times*, June 5, 2006.

Gorelick, S. "Hiding Damaging Information from the Public." *Ecologist* 28, no. 5 (1998): 301.

Gorman, J. "Did PCBs Save the Stripers? A Fish Story." *New York Times*, March 25, 2003.

Grafton, A., et al. "Polychlorinated Biphenyls (PCBs) in Fish Roe." *Journal of Young Investigators* 14 (2005).

Grand Jury, Northern District of Illinois. "Indictment of Calandra et al." U.S. District Court, October 1980.

Grasman, K., et al. "Organochlorine-Associated Immunosuppression in Prefledgling Caspian Terns and Herring Gulls from the Great Lakes: An Eco-epidemiological Study." *Environmental Health Perspectives* 113, no. 9 (September 2005).

Gray, K., et al. "In Utero Exposure to Background Levels of Polychlorinated Biphenyls and Cognitive Functioning among School-age Children." *American Journal of Epidemiology* 162 (2005).

"Great Lakes Are Exhaling Toxic Chemicals." *U.S. Water News*, October 2001.

Greider, W. *Who Will Tell the People?* New York: Simon & Schuster, 1992.

Griffith, E. *The Triumphs and Troubles of Theodore Swann*. Birmingham, AL: Black Belt Press, 1999.

Grigg, B. "DDT, PCBs Not Linked to Higher Rates of Breast Cancer, an Analysis of Five Northeast Studies Concludes." *National Institute of Environmental Health Sciences* online, May 15, 2001.

Griscom, A. "How Green Was the Gipper? A Look Back at Reagan's Environmental Record." Grist.com, June 10, 2004.

Gross, L. "Rachel's Daughter." *Sierra*, September/October 1999.

Grossman, R. "How Long Shall We Grovel?" POCLAD (Program on Corporations, Law and Democracy) online, April 4, 2001.

Guillette, L. J. Interview by Doug Hamilton. *Frontline*, PBS, November 1997.

Gunderson, D., et al. "Biomarker Response and Health of Polychlorinated Biphenyl- and Chlordane-Contaminated Paddlefish from the Ohio River Basin, USA." *Environmental Toxicology and Chemistry* 19, no. 9 (2000).

Guo, Y. L., et al. "Yucheng and Yusho: The Effects of Toxic Oil in Developing Humans in Asia." In *PCBs: Recent Advances in Environmental Toxicology and Health Effects*, eds. Robertson and Hansen.

Hale, L. "Organic Food, Exercise, but also PCBs." The Tyee, www.thetyee.ca, June 1, 2006.

Hamilton, Doug. "Fooling with Nature." *Frontline*, PBS, June 2, 1998.

Hand, E. "Roundup Is Killing off Amphibians, Ecologist Says." *St. Louis Post-Dispatch*, August 10, 2005.

Hardell, L., et al. "Increased Concentrations of Polychlorinated Biphenyls, Hexa-

chlorobenzene and Chlordanes in Mothers to Men with Testicular Cancer."
   Environmental Health Perspectives 111 (June 7, 2003).
———. "Concentrations of Polychlorinated Biphenyls in Blood and the Risk for
   Testicular Cancer." *International Journal of Andrology* 27 (2004): 282–90.
Harder, T., et al. "Investigations on Course and Outcome of Phocine Distemper
   Virus Infection in Harbour Seals (Phoca vitulina) Exposed to Polychlori-
   nated Biphenyls." *Zentralblatt fur Veterinarmedizin. Reihe B* (Journal of veteri-
   nary medicine. Series B) 39, no. 1 (February 1992): 19–31.
Harding, K., et al. "The 2002 European Seal Plague: Epidemiology and Popula-
   tion Consequences." *Ecology Letters* 5 (2002): 727–32.
Hauser, R., et al. "The Relationship Between Human Semen Parameters and
   Environmental Exposure to Polychlorinated Biphenyls and p,p'-DDE." *En-
   vironmental Health Perspectives* 111, no. 12 (May 19, 2003).
Hawkins, T. R. "Rereading *Silent Spring.*" *Environmental Health Perspectives* 102
   (June–July 1994): 6–7.
Hayes, D. "Earth Day 2000: End Global Warming." Foundation for Global
   Community, E-mail edition. No. 50A (March/April 2000).
Hayes, T., et al. "Pesticide Mixtures, Endocrine Disruption, and Amphibian
   Declines: Are We Underestimating the Impact?" *Environmental Health Per-
   spectives* 114, no. S-1 (April 2006).
Health Canada. "Fish and Seafood Survey." www.hc-sc.gc.ca/index-eng.php,
   2002.
Heilprin, J. "Class-action Lawsuit Being Filed against Dupont over Teflon
   Chemical Risks." Associated Press, July 20, 2005.
———. "Democrats Challenge EPA Pesticide Rule." Associated Press, October
   13, 2006.
Heinzerling, L., et al. "Amicus Brief—GE v. Whitman." U.S. District Court,
   April 2, 2001.
Hennig, B., et al. "PCBs and Cardiovascular Disease: Endothelial Cells as a Tar-
   get for PCB Toxicity." In *PCBs: Recent Advances in Environmental Toxicology and
   Health Effects*, eds. Robertson and Hansen.
Herman, R. "Prenatal Exposure to Mercury from a Maternal Diet High in Sea-
   food Can Irreversibly Impair Certain Brain Functions in Children." Harvard
   School of Public Health, news release. February 6, 2004.
———. "Exposure to PCBs May Reduce the Effectiveness of Vaccines in Chil-
   dren." Harvard School of Public Health, news release, August 22, 2006.
Higgs, S. "Dying for Their Work." CounterPunch online, March 4-5, 2006.
Hilderling, J. "Effecting Awareness of the Oceans We Share." PCB symposium
   in Malaysia, Japan Offspring Fund, 2003.
Hites, R. A. "The Integrated Atmospheric Deposition Network." *SPEA Insight*,
   www.spea.indiana.edu, 2006.

Hites, R., et al. "Global Assessment of Organic Contaminants in Farmed Salmon." *Science* 303 (January 9, 2004).

Holoubek, I. "Polychlorinated Biphenyl (PCB) Contaminated Sites Worldwide." In *PCBs: Recent Advances in Environmental Toxicology and Health Effects*, eds. Robertson and Hansen.

Hond, D. E., et al. "Sexual Maturation in Relation to Polychlorinated Aromatic Hydrocarbons: Sharpe and Skakkebaek's Hypothesis Revisited." *Environmental Health Perspectives* 110 (August 8, 2002).

Hooper, K. "Environmental Breast Milk Monitoring Programs (BMMPs)." *Environmental Health Perspectives* 107, no. 6 (June 1999).

Hooper, S. "Ecological Fate, Effects and Prospects for the Elimination of Environmental Polychlorinated Biphenyls (PCBs)." *Environmental Toxicology and Chemistry* 9, no. 5 (1990): 655–67.

Hosie, S., et al. "Is There a Correlation Between Organochlorine Compounds and Undescended Testes?" *European Journal of Pediatric Surgery* 10 (2000): 304–9.

Houlihan, J., et al. "BodyBurden: The Pollution in Newborns." Environmental Working Group, July 14, 2005.

Hunter, D., et al. "Plasma Organochlorine Levels and the Risk of Breast Cancer." *New England Journal of Medicine* 337 (1997).

Industrial Disease Standards Panel Canada. "Report to the Workers' Compensation Board on Occupational Exposure to PCBs and Various Cancers." IDSP report no. 2, December 1987.

International Program on Chemical Safety. *Polychlorinated Biphenyls and Terphenyls* (2nd edition). Environmental Health Criteria 140. Geneva: World Health Organization, 1993.

Jacobson, J. "Trends of Persistent Pollutants in Umbilical Cord Blood of Inuit Infants." *Environment and Health News*, April 17, 2006.

Jacobson, J. S. "Developmental Effects of PCBs in the Fish Eater Cohort Studies." In *PCBs: Recent Advances in Environmental Toxicology and Health Effects*, eds. Robertson and Hansen.

Jaenisch, et al. "Epigenetic Regulation of Gene Expression." Genetics supplement, *Nature* 33 (March 2003).

James, M. "Polychlorinated Biphenyls: Metabolism and Metabolites." In *PCBs: Recent Advances in Environmental Toxicology and Health Effects*, eds. Robertson and Hansen.

Janofsky, M. "Changes May Be Needed in Superfund, Chief Says." *New York Times*, December 5, 2004.

Javers, E. "Columnist Backed by Monsanto." *BusinessWeek*, January 13, 2006.

Jensen, S. "Report of a New Chemical Hazard." *New Scientist*, December 1966.

———. "The PCBs Story." *Ambio* 1(1972): 123–31.

Jha, A. "Man-made Chemicals Are Causing Serious Problems for Wild Animals." *Guardian*, May 29, 2003.

Jirtle, Randy. Interview ("Ask the Expert"). *Nova* ("Ghost in Your Genes"), PBS, August 2, 2007.

Johnson, Barry L., et al. "Public Health Implications of Exposure to Polychlorinated Biphenyls (PCBs)." ATSDR (Agency for Toxic Substances and Disease Registry) online, 2006.

Johnson, K. (2000). "GE Switches Its Web Site about PCBs." *New York Times*, September 30, 2000.

Joling, D. "Northern Fur Seal Pup Estimates Decline." Associated Press, February 5, 2007.

Josephson, J. "POP Surprise." *Environmental Health Perspectives* 110, no. 4 (April 2002).

Kalantzi, O.I. "The Global Distribution of PCBs and Organochlorine Pesticides in Butter." *Environmental Science & Technology* 35, no. 6 (2001).

Kaplan, S. "Kids at Risk: Chemicals in the Environment Come Under Scrutiny." *U.S. News & World Report*, June 19, 2000.

Karmaus, W., et al. "Parental Concentration of Dichlorodiphenyl Dichloroethene and PCBs in Michigan Fish Eaters and Sex Ratio in Offspring." *Journal of Occupational Medicine* 44 (January 1, 2002).

Kay, J. "Sick, Dying Sea Otters Turn Up in Morro Bay." *San Francisco Chronicle*, April 25, 2004.

Keller, J., et al. "Organochlorine Contaminants in Loggerhead Sea Turtle Blood: Extraction Techniques and Distribution among Plasma and Red Blood Cells." *Archives of Environmental Contamination and Toxicology* 46 (2004): 254–64.

Kennedy, S., et al. "Mass Die-Off of Caspian Seals Caused by Canine Distemper Virus." *CDC* 6, no. 6 (November/December 2000).

Kim, K. H., et al. "Excitatory and Inhibitory Synaptic Transmission Is Differentially Influenced by Two Ortho-Substituted Polychlorinated Biphenyls in the Hippocampal Slice Preparation." *Toxicology and Applied Pharmacology* 237, no. 2 (2009): 68–77.

Kimbrell, A. "The Frankenstein Corporation: Monsanto's Merger with American Home." *Ecologist* 28, no. 5 (1998): 306–8.

Kingsnorth, P. "Bovine Growth Hormones." *Ecologist* 28, no. 5 (1998): 266–69.

Kinney, J. "Dredging Must Wait until 2008." *Saratogian*, July 28, 2006.

Klotz, L. "Why Is the Rate of Testicular Cancer Increasing?" *Canadian Medical Association Journal* 160, no. 2 (January 26, 1999): 213–14.

Kodavanti, P., et al. "Differential Effects of Two Lots of Aroclor 1254: Congener-Specific Analysis and Neurochemical End Points." *Environmental Health Perspectives* 109, no. 11 (November 2001).

Kolata, Gina. "Study Discounts DDT Role in Breast Cancer." *New York Times*, October 30, 1997.

Koopman-Esseboom, C., et al. "Effects of Polychlorinated Biphenyl/Dioxin Exposure and Feeding Type on Infants' Mental and Psychomotor Development." *Pediatrics* 97, no. 5 (1996): 700–706.

Korrick, S. A. (2001). Polychlorinated Biphenyls (PCBs) and Neurodevelopment in General Population Samples. In *PCBs: Recent Advances in Environmental Toxicology and Health Effects*, eds. Robertson and Hansen.

Krimsky, S. *Hormonal Chaos*. Baltimore, MD: Johns Hopkins University Press, 2000.

Kuhn, W. "Aroclor: Wildlife." Monsanto interoffice memo, Monsanto company records, December 30, 1968. "The Inside Story," Environmental Working Group, www.ewg.org.

Kurzweil, R. *The Singularity Is Near*. New York: Viking, 2005.

Laden, F., et al. "1,1-Dichloro-2,2-bis(p-chlorophenyl)ethylene and Polychlorinated Biphenyls and Breast Cancer: Combined Analysis of Five U.S. Studies." *Journal of the National Cancer Institute* 93, no. 10 (May 16, 2001): 768-76.

Laden, F., et al. "Polychlorinated Biphenyls, Cytochrome P450 1A1, and Breast Cancer Risk in the Nurses' Health Study." *Cancer Epidemiology* 11 (December 2002): 1560–65.

Lahvis, G., et al. "Decreased Lymphocyte Responses in Free-ranging Bottlenose Dolphins (Tursiops truncatus) Are Associated with Increased Concentrations of PCBs and DDT in Peripheral Blood." *Environmental Health Perspectives* 103, suppl. 4 (May 1995).

Lambsound. "John F. Queeny and Monsanto." Lambsound Starting Point, www.lambsound.com. April 19, 2006.

Landy, M. K. *The Environmental Protection Agency*. New York: Oxford University Press, 1994.

"Latest Dredging Delay Unfair to the River" (editorial). *Record* (Troy, NY), August 3, 2006.

LeBoeuf, B., et al. "Organochloride Pesticides in California Sea Lions Revisited." *BioMedical Central Ecology* 2, no. 1 (2002): 11.

Lederburg, J., et al. "'Ome Sweet 'Omics." *Scientist* 15, no. 8 (April 2, 2001).

Lee, J. "Judge Rules Sea Lion Research Violates Laws." Associated Press, May 31, 2006.

Lehner, P. "Setting the Record Straight about Legality of G.E.'s PCB Discharges." HudsonWatch.net, January 5, 2001.

Lemonick, M. "Teens before Their Time." *Time*, October 30, 2000.

Leney, J., et al. "Does Metamorphosis Increase the Susceptibility of Frogs to

Highly Hydrophobic Contaminants?" *Environmental Science and Technology* 40 (2006).

"Letters from GE and Monsanto." *Sierra*, November/December 1994.

Li, Y., et al. "Polychlorinated Biphenyls, Cytochrome P450 1A1 (CYP1A1) Polymorphisms, and Breast Cancer Risk Among African American Women and White Women in North Carolina: A Population-Based Case-Control Study." *Breast Cancer Research* 7, no. 1 (2004).

Little, A. G. "G.E.'s Green Gamble." *Vanity Fair*, July 2006.

Lohman, R. "PCBs Reaching the Deep Oceans." *Geophysical Research Letters*, October 17, 2006.

Lomborg, B. *The Skeptical Environmentalist* (Cambridge, UK: Cambridge University Press, 2001).

Longnecker, M. "Endocrine and Other Human Health Effects of Environmental and Dietary Exposure to Polychlorinated Biphenyls." In *PCBs: Recent Advances in Environmental Toxicology and Health Effects*, eds. Robertson and Hansen.

Loof, S. "Ukrainian Presidential Candidate Viktor Yushchenko Poisoned with Dioxin." Associated Press, December 11, 2004.

Loomis, D., et al. "Cancer Mortality Among Electric Utility Workers Exposed to Polychlorinated Biphenyls." *Occupational and Environmental Medicine* 54 (1997).

Lortie, J., et al. "Monitoring Amphibian Reproductive Success at a PCB-Contaminated Site Using Fluorescent Pigments." Poster presented at Society for Environmental Toxicology and Chemistry 23rd annual meeting, Salt Lake City, UT, 2002.

Love, D. *My City Is Gone* (New York: William Morrow, 2006).

Lovelock, J. E. *Gaia: A New Look at Life on Earth*. Oxford, UK: Oxford University Press, 1979.

———. *Homage to Gaia*. Oxford, UK: Oxford University Press, 2000.

Lowenstein, L. "Sick Sea Mammals: A Sign of Sick Seas?" International Veterinary Information Service Web site, November 13, 2004.

Lowy, J. "Autism Reaching 'Epidemic' Levels." Scripps Howard News Service, January 21, 2004.

Ludwig, J., et al. "A Comparison of Water Quality Criteria for the Great Lakes Based on Human and Wildlife Health." *Journal of Great Lakes Research* 19, no. 4 (1993).

Luebke, R., et al. "Aquatic Pollution-Induced Immunotoxicity in Wildlife Species" (abstract). *Fundamental and Applied Toxicology* 37, no. 1 (May 1997).

Lymphoma Net. "Exposure to Polychlorinated Biphenyls May Increase NHL Risk." January 27, 2006.

MacGillis, D. "At the Toxic Trade Show." *Boston Globe*, November 27, 2006.

MacInnis, L. "U.S. Study Links Chemical to Sperm Damage." *USA Today*, December 11, 2002.

Mackenzie, D. "Eels Slide Toward Extinction." *New Scientist. 180,* no. 2415 (October 4, 2003).

Magnuson, E. "A Problem That Cannot Be Buried: The Poisoning of America Continues." *Time,* October 14, 1985.

Malawa, Z. "The Cancer Cow." *Nutrition Bytes* 8, no. 1 (2002) article 4.

"Malignant Mimicry" (editorial). *Scientific American,* September 1993.

Malkin, M. "Rachel's Folly: The End of Chlorine." Competitive Enterprise Institute, March 1996.

Manchester-Neesvig, J. B., et al. "Comparison of Polybrominated Diphenyl Ethers (PBDEs) and Polychlorinated Biphenyls (PCBs) in Lake Michigan Salmonids." *Environmental Science and Technology* 35, no. 6 (2001).

Marcotty, J. "Fish Is Health, but Not Always." *Minneapolis–St. Paul Star Tribune,* December 17, 2006.

Martindale, D. Letter to the editor. *Berkshire* (Mass.) *Eagle,* April 30, 2006.

Matheson, D. "Dead Killer Whale Raises Concerns about Pollution." CTV News and Current Affairs online, March 21, 2000.

McCaffrey, S. "GE Files 'Good Faith Offer' with EPA for Plan to Clean Up Hudson River." Associated Press, April 8, 2002.

McClure, R. "Dead Orca Is a 'Red Alert.'" *Seattle Post-Intelligencer,* May 7 2002.

———. "Sound's Salmon Carry High PCB Levels." *Seattle Post-Intelligencer,* January 15, 2004.

McElroy, J., et al. "Potential Exposure to PCBs, DDT, and PBDEs from Sport-Caught Fish Consumption in Relation to Breast Cancer Risk in Wisconsin." *Environmental Health Perspectives* 112, no. 2 (February 2, 2004).

McGrane, J. "Environmentalists Urge Rejection of GE's PCB Cleanup Offer." Media release, Friends of a Clean Hudson, April 12, 2006.

McKibben, B. *The End of Nature.* New York: Random House, 1986.

McKinley, J. "Heading to Texas, Hudson's Toxic Mud Stirs Town." *New York Times,* May 30, 2009.

Meander, G. "Chytrid Fungus Implicated as Factor in Decline of Arizona Frogs." U.S. Geological Service, March 29, 2000.

Medical Science News. "Natural Organohalogens Have Potent Anticancer and Antibacterial Activity." *Medical Science News,* July 22, 2004.

Mele, A. "River Truth." Hudson River Sloop/Clearwater, Clearwater.org, April 11, 1998.

Mendola, P., et al. "Birth Defects Risk Associated with Maternal Sport Fish Consumption: Potential Effect Modification by Sex of Offspring." *Environmental Research* 10 (2003).

Messick, G. "Toxic Town." *60 Minutes,* CBS, August 31, 2003.

Metrick, A. "NRDC Calls on Whitman to Move Speedily on PCB Clean-Up Plan." Natural Resources Defense Council, news release, October 4, 2001.

Miao, X., et al. "Congener-specific Profile and Toxicity Assessment of PCBs in Green Turtles (Chelonia mydas) from the Hawaiian Islands." *Science of the Total Environment* 281, nos. 1–3 (December 17, 2001).

Milloy, S. "Lame Duck EPA Administrator Carol Browner." Fox News.com, December 8, 2000.

———. "Final Countdown at EPA." JunkScience.com, December 8, 2000.

———. "When Environmental and Political Science Clash." JunkScience.com, December 7, 2001.

———. "Fishy Dietary Advice." Fox News.com, October 19, 2006.

Mills, M. "2000 Ohio River Fish Consumption Advisory." Kentucky Division of Water, www.water.ky.gov, March 20, 2006.

M.I.N.D. Institute. "Autism News." University of California–Davis archives, May 6, 2005.

Minners, J. "City Removes Squatters, Debris from Toxic Hexagon Site." *Bronx News*, December 5, 2002.

Mittelstaedt, M. "The New PCBs." (Toronto) *Globe and Mail*, June 5, 2004.

———. "Flame Retardant in Breast Milk Raises Concern." *Globe and Mail*, June 7, 2004.

———. "Toxic Cocktail Found in Children." *Globe and Mail*, February 6, 2006.

Mokhiber, R., et al. "Pulp Non-Fiction: The Ecologist Shredded." In *Corporate Predators: The Hunt For Mega-Profits and the Attack on Democracy*. Monroe, ME: Common Courage Press, 2002.

Moller, H. "Trends in Sex-Ratio, Testicular Cancer and Male Reproductive Hazards: Are They Connected?" *Acta Pathologica, Microbiologica et Immunologica Scandinavica* 106 (1998).

Monroe, D. "Amphibians Suffering Unprecedented Decline, Global Study Finds." *Scientific American*, October 15, 2004.

Monsanto company records. Allen, J. H. Letter to Dr. Emmet Kelly. February 14, 1961. "The Inside Story," Environmental Working Group, www.ewg.org.

———. "Salesmen's Manual Aroclor, Description and Properties." October 1, 1944. "The Inside Story," Environmental Working Group, ewg.org.

———. "Monsanto Europe letter to Monsanto US re: Aroclor Sweden." December 1, 1966. "The Inside Story," Environmental Working Group, www.ewg.org.

———. "Monsanto Memo: PCB Preparedness Q&A." September 29, 1976. "The Inside Story," Environmental Working Group, www.ewg.org.

———. "RE: Histopathological Evaluation." J. C. Calandra letter to Monsanto. March 24, 1975. "The Inside Story," Environmental Working Group, www.ewg.org.

———. Industrial Bio-Test Laboratories. "Review of PCB Meeting." April 15, 1975. "The Inside Story," Environmental Working Group, www.ewg.org.

————. "Biomedical and Environmental Special Programs." November 2, 1982. "The Inside Report," Environmental Working Group, www.ewg.org.

————. Hamer W. "Aroclor Toxicity" (letter). February 27, 1954. "The Inside Story," Environmental Working Group, www.ewg.org.

————. "Dermatitis Report: Anniston." Monsanto interoffice memo. May 1935. "The Inside Story," Environmental Working Group, www.ewg.org.

————. "Report to the Monsanto Chemical Company" (Report 4465). September 15, 1938. "The Inside Story," Environmental Working Group, www.ewg.org.

————. "Process for the Production of Aroclors etc. at the Anniston and Krummrich Plants" (E. Mather letter). April 1955. "The Inside Story," Environmental Working Group, www.ewg.org.

————. Gordon, D. "Kimbrough Study." March 24, 1975. "The Inside Story," Environmental Working Group, www.ewg.org.

————. "Final Report on Aroclor and Gases." A. Ellenburg. March 15, 1954. "The Inside Story," Environmental Working Group, www.ewg.org.

————. "Department 246: Aroclors" (J. M. Garrett letter). November 14, 1955. "The Inside Story," Environmental Working Group, www.ewg.org.

————. "Aroclors Application Information" (E. M. Kelly letter). February 14, 1950. "The Inside Story," Environmental Working Group, www.ewg.org.

————. "Maximum Allowable Concentration of Aroclor" (letter). February 12, 1954. "The Inside Story," Environmental Working Group, www.ewg.org.

————. "Aroclor Toxicity" (letter). September 20, 1955. "The Inside Story," Environmental Working Group, www.ewg.org.

————. "Pydraul" (letter). June 7, 1956. "The Inside Story," Environmental Working Group, www.ewg.org.

————. "Pydraul 150" (letter). January 21, 1957. "The Inside Story," Environmental Working Group, www.ewg.org.

————. "Memo." June 23, 1959. "The Inside Story," Environmental Working Group, www.ewg.org.

————. "Hexagon Labs." February 2, 1961. "The Inside Story," Environmental Working Group, www.ewg.org.

————. Levinskas, G. "Aroclor 2-Year Rat Feeding Studies." July 18, 1975. "The Inside Story," Environmental Working Group, www.ewg.org.

————. "Confirming our phone conversation." Letter to Marcus Key, MD, March 15, 1962. "The Inside Story," Environmental Working Group, www.ewg.org.

————. "Aroclor Sweden." Monsanto company records, December 12, 1966. "The Inside Story," Environmental Working Group, www.ewg.org.

————. Flinn, F. B. "Report: Animal Research Test Results." Monsanto interoffice memo, May 25, 1934. "The Inside Story," Environmental Working Group, www.ewg.org.

————. Widmark, Gunnar. "Visit to Monsanto—Thanks" (letter). December 29, 1966. "The Inside Story," Environmental Working Group, www.ewg .org.

————. Smith, D. "Pydraul Labels." December 5, 1958. "The Inside Story," Environmental Working Group, www.ewg.org.

————. Olson, D. "PCB Toxicity Problem Proper Disposal of Scrap Aroclor." Monsanto interoffice memo, December 5, 1969. "Inside Story," Environmental Working Group, www.ewg.org.

————. Wheeler, E. M. "Aroclors Cannot Be Considered to Be Non-toxic." September 1, 1953. "The Inside Story," Environmental Working Group, www. ewg.org.

————. "Chloracne Cases." June 12, 1956. "The Inside Story," Environmental Working Group, www.ewg.org.

————. "Pydraul 150." December 26, 1956. "The Inside Story," Environmental Working Group, www.ewg.org.

————. "Toxicity—Pydraul 150." September 25, 1957. "The Inside Story," Environmental Working Group, www.ewg.org.

————. "Aroclor 1232." May 27, 1964. "The Inside Story," Environmental Working Group, www.ewg.org.

————. "Aroclor 1242." September 3, 1965. "The Inside Story," Environmental Working Group, www.ewg.org.

————. Strand, H. "Aroclors: Jensen." November 28, 1966. "The Inside Story," Environmental Working Group, www.ewg.org.

————. Scientific Associates. "Certificate of Analysis." November 10, 1953. "The Inside Story," Environmental Working Group, www.ewg.org.

"Monsanto Products Used in World War II." Monsanto company records. 1945. Series 10 (box 4), archives, Washington University, St. Louis.

Montague, Peter. "Waste Management Inc. Sues EPA to Force Agency to Issue Rules Sanctioning Ocean Incineration." *Rachel's Democracy & Health News* 14 (March 1, 1987).

————. "EPA Names 100 Most Hazardous Chemicals at Superfund Dumps." *Rachel's Democracy & Health News* 27 (May 31, 1987).

————. "Waste Haulers Discover a Place to Build Dumps and Incinerators Where Regulations Are Often Lax." *Rachel's Democracy & Health News* 48 (October 25, 1987).

————. "Environmental Groups Charge EPA Is Ignoring Federal Superfund Law." *Rachel's Democracy & Health News* 56 (December 21, 1987).

————. "More Lessons from Superfund (Part 2)." *Rachel's Democracy & Health News* 87 (July 25, 1988).

————. "Human Breast Milk Is Contaminated." *Rachel's Democracy & Health News* 193 (August 7, 1990).

——. "New Study Links Breast Cancer to DDT, PCBs." *Rachel's Democracy & Health News* 279 (April 1, 1992).

——. "Push Comes to Shove." *Rachel's Democracy & Health News* 234 (May 21, 1991).

——. "As the PCBs Story Unfolds." *Rachel's Democracy & Health News* 295 (July 22, 1992).

——. "International Waste Trade: Part 1; A Flaw in the Grass-Roots Strategy." *Rachel's Democracy & Health News* 253 (October 3, 1991).

——. "New EPA Memo Says All Hazardous Waste Incinerators Fail to Meet Regulations." *Rachel's Democracy & Health News* 312 (November 17, 1992).

——. "Fixing Superfund: Part 1." *Rachel's Democracy & Health News* 357 (September 29, 1993).

——. "Chemicals and Health: Part 1." *Rachel's Democracy & Health News* 369 (December 23, 1993).

——. "Bruce Ames." *Rachel's Democracy & Health News* 398 (July 14, 1994).

——. "Dangers of Chemical Combinations." *Rachel's Democracy & Health News* 498 (June 13, 1996).

——. "PCBs: Smaller Penis." *Rachel's Democracy & Health News* 372 (August 10, 1996).

——. "PCB Exposure Linked to Low IQ." *Rachel's Democracy & Health News* 512 (September 18, 1996).

——. "Chemicals Linked to Declining Male Reproductive Health." *Rachel's Democracy & Health News* 514 (October 3, 1996).

——. "Frogs Give Warning." *Rachel's Democracy & Health News* 515 (October 10, 1996).

——. "Shifting the Burden of Proof." *Rachel's Democracy & Health News* 491 (April 25, 1996).

——. "Infectious Disease and Pollution." *Rachel's Democracy & Health News* 528 (January 9, 1997).

——. "Statement on Immune Toxins." *Rachel's Democracy & Health News* 536 (March 6, 1997).

——. "The Causes of Lymph Cancers." *Rachel's Democracy & Health News* 562 (September 4, 1997).

——. "The Toxic Substances Control Act." *Rachel's Democracy & Health News* 564 (September 18, 1997).

——. "The Truth about Breast Cancer: Part 4." *Rachel's Democracy & Health News* 574 (November 27, 1997).

——. "Follow the Money." *Rachel's Democracy & Health News* 307 (February 25, 1998).

——. "A Campaign of Reassuring Falsehoods." *Rachel's Democracy & Health News* 656 (June 23, 1999).

————. "Torching the Environment." *Rachel's Hazardous Waste News*, 1992. www
.multinationalmonitor.org/hyper/issues/1992/05/mm0592_05.html.

————. "How Monsanto Listens." *Ecologist* 28, no. 5 (1998).

Moore, C. "Trashed: Across the Pacific Ocean, Plastics, Plastics, Everywhere."
Natural History 112, no. 9 (November 2003).

————. *Silent Scourge*. New York: Oxford University Press, 2003.

Morgan, J. "Joe Mantegna Rules in Favor of Autism Funding." *USA Today*, January 14, 2002.

Morris, F. "Monsanto: You Have Shamed Us." *Ecologist* 28, no. 5 (1998).

"Mothers' PCBs and Sons' Testicular Cancer." Our Stolen Future Web site, www
.ourstolenfuture.org/.

Moysich, K. B. "Environmental Exposure to Polychlorinated Biphenyls and
Breast Cancer Risk." In *PCBs: Recent Advances in Environmental Toxicology and
Health Effects*, eds. Robertson and Hansen.

————. "Research Commentary: Thoughts on Recent Finds Regarding Organochlorines and Breast Cancer Risk." *Ribbon* 6, no. 3 (fall 2001).

Mund, C. et al. "Reactivation of Epigenetically Silenced Genes by DNA Methyl-
transferase Inhibitors." *Epigenetics* 1, no. 1 (January–March 2006).

Murugesan, P., et al. "Studies on the Protective Role of Vitamin C and E against
Polychlorinated Biphenyl (Aroclor 1254)–Induced Oxidative Damage in Ley-
dig Cells." *Free Radical Research* 39, no. 11 (2005).

Myers, J. P. "The Next Victim in Bush's War on the Environment." *New Jersey
Star Ledger*, April 11, 2001.

Nash, M. "The Secrets of Autism." *Time*, April 29, 2002.

National Cancer Institute. "DES: Questions and Answers." National Cancer
Institute online, April 15, 2003.

National Institute of Neurological Disorders and Stroke, National Institutes of
Health. *Parkinson's Disease: Challenges, Progress, and Promise*. NIH publication
no. 05-5595. December 2004.

National Research Council. *Hormonally Active Agents in the Environment*. Washington, D.C.: National Academies Press, 1999.

Natural Resources Defense Council. "Healing the Hudson." August 31, 2002.

————. "EPA Lifts Ban on Selling Polluted Sites for Development." September
2, 2003.

————. "Hudson River Data Report: Contaminants in Bullfrog and Snapping
Turtle." February 28, 2005.

————. "Healthy Milk, Healthy Baby: Chemical Pollution and Mother's Milk."
March 25, 2005.

————. "Our Children at Risk." www.nrdc.org/health/kids/ocar/ocarinx.asp,
November 1997.

Neiwert, D. "The First Ted Olson Scandal." Salon.com, May 14, 2001.

Ness, E. "Dirty Minds: Colleen Moore's Silent Scourge Tallies the Toll of Pollution on Kids' Brains." Grist.com, December 15, 2003.

Newbold, R., et al. "Adverse Effects of the Model Environmental Estrogen Diethylstilbestrol Are Transmitted to Subsequent Generations." *Endocrinology* 147, no. 6 (2006).

Nielsen, E., et al. "Children and the Unborn Child: Exposure and Susceptibility to Chemical Substances—An Evaluation." *Miloprojekt* no. 589 (2001).

"No Link Seen Between Breast Cancer and Pesticides, PCB Exposure for General Population." University of Buffalo news release, www.scienceblog.com/community/older/1997/B/199701318.html, November 1, 1997.

Novak, R. "The Long Arm of the Lab Laws." TCAW (Testicular Cancer Awareness Week), www.tcaw.org.

Novick, S., and D. Cotrell, eds. *Our World in Peril.* Greenwich, CT: Fawcett, 1971.

Novovitch, B. "Texas Falcon Tracker Finds Foul Environment." Reuters, October 15, 2004.

Nowell, L. H. "Organochlorine Pesticides and PCBs in Bed Sediment and Whole Fish from United States Rivers and Streams." *National Water Quality Assessment Program, 1992–2001,* 2003.

O'Boyle, T. F. *At Any Cost.* New York: Alfred A. Knopf, 1998.

Oceana. "Toxic Burden: PCBs in Marine Life." Oceana.org, 2006.

Oceans Alive! "Consumption Advisories: Fish to Avoid." Oceans Alive online, 2006.

Oddi, M. "More on D.C. Circuit's CERCLA Ruling This Week." Indiana Law Blog, www.indianalawblog.com, March 5, 2004.

O'Harra, D. "Whales in Sound Imperiled." *Anchorage Daily News,* July 22, 2001.

Opanashuk, L., et al. "PCBs, Fungicide Open Brain Cells to Parkinson's Assault." *University of Rochester Medical Center* news online, January 26, 2005.

"An Open Letter to Robert Shapiro, Chief Executive Officer of Monsanto" (editorial). *Ecologist* 28, no. 5 (1998).

O'Shea, T. J. "Marine Mammals and Persistent Ocean Contaminants." Proceedings of the Marine Mammal Commission Workshop, Keystone, Colorado, October 12–15, 1998.

———. "Cause of Seal Die-off in 1988 Is Still under Debate" (letters). *Science* 290, nos. 5494, 1097 (2000).

Owen, J. "Europe's Eels Are Slipping Away, Scientists Warn." *National Geographic News,* October 9, 2003.

Pacenza, M. "PCB Dredging Reaffirmed." *Times Union* (Albany, NY), May 20, 2005.

———. "Latest Maneuvers Appear Certain to Delay Dredging" *Times Union,* July 19, 2006.

Palstra, A., et al. "Are Dioxin-Like Contaminants Responsible for Eel (Anguilla Anguilla) Drama?" *Naturwissenschaften* 93 (2006).

———. "Eels Extinct from Dioxins." Leiden University news archive. www .leidenuniv.nl/en/researcharchive, March 2006.

Parfitt, B. "Poison Evidence Points to Illegal Dumping." Creative Resistance online, www.creativeresistance.ca, November 25, 2002.

Parrent, K. "Environmentalists Applaud EPA Hudson River Clean-Up Decision." National Resource Defense Council, news release, August 1, 2001.

———. "Environmentalists Fight General Electric's Latest Evasion of PCB Cleanup." Scenic Hudson, news release, April 6, 2001.

"PCBs, Fungicide Open Brain Cells to Parkinson's Assault." University of Rochester Medical Center online, January 31, 2005.

Pearce, F. "'Miscalculation' Could Mean the End of Caviar." *New Scientist*, September 17, 2003.

Pellegrini, F. "Bush vs. Big Business? You Never Know." *Time*, August 1, 2001.

Peper, M., et al. "Neuropsychological Effects of Chronic Low-Dose Exposure to Polychlorinated Biphenyls (PCBs): A Cross-Sectional Study." *Environmental Health*, no. 4 (2005).

Perlman, D. "Bay Scientist's Pollutant Warning." *San Francisco Chronicle*, February 24, 1969.

———. "The Effects of PCB Exposure and Fish Consumption on Endogenous Hormones." *Environmental Health Perspectives* 109, no. 12 (December 2001).

Persky, V. W. "Health Effects of Occupational Exposure to PCBs." In *PCBs: Recent Advances in Environmental Toxicology and Health Effects*, eds. Robertson and Hansen.

Pessah, I. "Etiology of PCB Neurotoxicity: From Molecules to Cellular Dysfunction." In *PCBs: Recent Advances in Environmental Toxicology and Health Effects*, eds. Robertson and Hansen.

Pessah, I., et al. "A New Generation of Environmentally Relevant PCBs: Structure-Activity, Molecular Mechanisms, and Developmental Neurotoxicity." University of California-Davis news release, May 12, 2006.

Petersen, R. "Appellate Court Allows Constitutional Challenge to Superfund Provision to Go Forward." *Environmental Law Update*, July/August 2004.

Pfeiffer, B. "The Fate of Frogs." Vermont Public Interest Research Group (VPIRG) online, www.vpirg.org, October 1999.

Pianin, E. "Former N.J. Governor's Tenure Gets Mixed Reviews." *Washington Post*, May 22, 2003.

Pierce, G., et al. "Pollutants in the Sea—A Threat to Dolphin and Porpoise Populations?" University of Aberdeen news release, June 14, 2004.

Plapp, F. "Environmental Chemicals and Environmental Illness: A Major Role for Vitamin A." Texas A&M University online archives, 2006.

Pollak, R. "Is GE Mightier Than the Hudson?" *Nation*, May 28, 2001.

Pope, C. "Sierra Club's Reaction to Whitman Resignation." Truthout.org, May 21, 2003.

Powers, M. "Deadly Fungus Threatens Beloved Panamanian Frog." Reuters, February 7, 2006.

Pray, L. "Epigenetics: Genome, Meet Your Environment." *Scientist* 18, nos. 13/14, July 5, 2004.

Prichard, J. "EPA Chief Christie Whitman Announces Plan to Clean Up and Restore the Great Lakes." Associated Press, April 2, 2002.

"Profile: The Monsanto Company." St. Louis Historic Preservation Web site, 2006.

Queeny, E. M. *The Spirit of Enterprise*. New York: Charles Scribner's Sons, 1943.

"Rachel Carson Dies of Cancer" (editorial). *New York Times*, April 15, 1964.

Raloff, J. "Those Old Dioxin Blues." *Science News*, May 17, 1997.

———. "Pollution Fallout." *Science News* 171, no. 13 (March 31, 2007).

Ray, D. L. *Environmental Overkill*. Washington, D.C.: Regnery Gateway Books, 1993.

Rayl, A. "Researching Heavy Metal Contamination in Arctic Whales." *Scientist* 13, no. 13 (November 22, 1999).

Reeder, A., et al. "Forms and Prevalence of Intersexuality and Effects of Environmental Contaminants on Sexuality in Cricket Frogs (Acris crepitans)." *Environmental Health Perspectives* 106, no. 5 (May 1998).

Reformation Online. "Genetically Modified Food." Reformation.org. September 14, 2002.

Regenstein, L. *America the Poisoned*. New York: Acropolis Books, 1982.

Rehmeyer, J. J. "Salmon Safety." *Science News*, January 20, 2007.

Reiss, W. P. "GE Says Will Clean-up and Sues." Friends of a Clean Hudson, media release, February 15, 2002.

Rendon, J. "Puget Sound's Alarm: Banned PCBs Still Threaten Marine Life." *Sierra*, July/Aug 2001.

Renner, R. "NIEHS-Funded Research Pursues Thyroid Findings." *Environmental Health Perspectives* 111, no. 12 (September 2003).

Reuters. "Hudson River PCBs Seeping from Sediment." July 24, 1998.

———. "Microbes Can Clean Up Toxic Waste Dumps, Scientist Says." September 8, 2006.

Richards, J. "Inherited Epigenetic Variation." *Nature Reviews Genetics AOP*, doi: 10.1038/nrg1834, March 14, 2006.

Risebrough, R. "Endocrine Disrupters and Bald Eagles: A Response." www.umich.edu/~esupdate/library/98.05-06/risebrough.html.

———. "'Endocrine Disruption' and the Wildlife Connection." *Human and Ecological Risk Assessment* 5 (1999).

Risebrough, R., et al. "Polychlorinated Biphenyls in the Global Ecosystem." *Nature* 220 (December 14, 1998).

Riverkeeper. "Endorsement Letter." Riverkeeper.org, April 6, 2001. ·

Rizzo, M. "PCB Exposure May Raise Lymphoma Risk." Reuters, June 30, 2005.

Roan, S. "Living for Two." *Los Angeles Times*, November 12, 2007.

Robertson, L. W., and L. G. Hansen, eds. *PCBs: Recent Advances in Environmental Toxicology and Health Effects.* Lexington, KY: University Press of Kentucky, 2001.

Robinet, T. "Sublethal Effects of Exposure to Chemical Compounds: A Cause for the Decline in Atlantic Eels?" *Ecotoxicology* 11 (2002).

Rose, D. "Children of Dads over 40 at Risk of Autism." *Australian*, September 5, 2006.

Rosen, R. "Polluted Bodies." *San Francisco Chronicle*, February 3, 2003.

Rosenblum, M. "Scientists Warn of Undetected, Unmeasured Toxins in World's Fish." Associated Press, November 18, 2004.

Rosenthal, E. "British Rethinking Rules after Ill-Fated Drug Trial." *New York Times*, April 8, 2006.

Ross, P., et al. "Contaminant-Related Suppression of Delayed-type Hypersensitivity and Antibody Responses in Harbor Seals Fed Herring from the Baltic Sea." *Environmental Health Perspectives* 103, no. 2 (February 1995).

———. "PCBs Are a Health Risk for Humans and Wildlife." *Science* 289 (September 15, 2000).

———. "High PCB Concentrations in Free-Ranging Pacific Killer Whales, Orcinus Orca: Effects of Age, Sex, and Dietary Preference." *Marine Pollution Bulletin* 40 (2001).

Rothman, N., et al. "A Nested Case-Control Study of Non-Hodgkin Lymphoma." *Lancet* 350 (1997).

Rowell, A. "SLAPPing the Resistance." *Ecologist* 28, no. 5 (1998).

Rozell, N. "Alaska's Fish Are Very Clean." *Alaska Report*, August 29, 2006.

Ruch, J. Letter to President Obama from PEER (Public Employees for Environmental Responsibility), December 4, 2008.

Ruder, A. M., et al. "Mortality among Workers Exposed to Polychlorinated Biphenyls (PCBs) in an Electrical Capacitor Manufacturing Plant in Indiana: An Update." *Environmental Health Perspectives*, September 1, 2005, doi: 10.1289/ehp.8253.

Rusiecki, J. A., et al. "A Correlation Study of Organochlorine Levels in Serum, Breast Adipose Tissue, and Gluteal Adipose Tissue among Breast Cancer Cases in India." *Cancer Epidemiology, Biomarkers & Prevention* 14, no. 5 (May 2005).

Ruzzin, J., et al. "Persistent Organic Pollutant Exposure Leads to Insulin Resistance." *Environmental Health Perspectives*, 118, no. 4 (April 2010).

Safe, S. "Endocrine Disruptors and Human Health: Is There a Problem? An Update." *Environmental Health Perspectives* 108, no. 6 (June 2000).

———. "Endocrine Disruptors and Pharmaceutically Active Compounds in Drinking Water Workshop." Center for Health Effects of Environmental Contamination, 2000 Annual Report, www.cheec.uiowa.edu/annreport_pdf.

———. "PCBs as Aryl Hydrocarbon Receptor Agonists: Implications for Risk Assessment." In *PCBs: Recent Advances in Environmental Toxicology and Health Effects*, eds. Robertson and Hansen.

Sang, S. "Assessment of Contaminants in Beluga Whales and Polar Bears Reproductive Systems." World Wildlife Fund-Canada, December 9, 2007.

Schantz, S. L. "Developmental Neurotoxicity of PCBs in Humans: What Do We Know and Where Do We Go from Here?" *Neurotoxicology and Teratology* 18 (May/June 1997).

Schantz, S. L., et al. "Effects of PCB Exposure on Neuropsychological Function in Children." *Environmental Health Perspectives* 111, no. 3 (March 2003).

Schecter, A. J., et al. "PCBs, Dioxins, and Dibenzofurans: Measured Levels and Toxic Equivalents in Blood, Milk and Food from Various Countries." In *PCBs:Recent Advances in Environmental Toxicology and Health Effects*, eds. Robertson and Hansen.

Schettler, T., et al. "In Harm's Way: Toxic Threats to Child Development." Greater Boston Physicians for Social Responsibility, May 2000.

Schmidt, C. "Poisoning Young Minds." *Environmental Health Perspectives* 107, no. 6 (June 6, 1999).

Schmitt, C. "Persistent Environmental Contaminants in Fish and Wildlife." National Biological Service, 2006.

Schneider, K. "Faking It." *Amicus Journal* (spring 1993).

Schwartz, J. "Amid Devastation, Mounds of Toxic Waste Vie for Attention." *New York Times*, December 20, 2005.

Seegal, R. "The Neurotoxicological Consequences of Developmental Exposure to PCBs." *Toxicological Sciences* 57, nos. 1–3 (2000).

Seelye, K. Q. "G.E. Is Accused of Trying to Undercut Order to Dredge Hudson River." *New York Times*, October 1, 2001.

Shapley, D. "Dredging project spurs precautions." *Poughkeepsie* (New York) *Journal*, May 12, 2006.

Sharpe, R. M. "Hormones and Testis Development and the Possible Adverse Effects of Environmental Chemicals." *Toxicology Letters* 120 (2001).

———. "Phthalate Exposure during Pregnancy and Lower Anogenital Index in Boys: Wider Implications for the General Population?" *Environmental Health Perspectives* 113, no. 8 (2005).

Sheehan, D., et al. "No Threshold Dose for Estradiol-Induced Sex Reversal of

Turtle Embryos: How Little Is Too Much?" *Environmental Health Perspectives* 107, no. 2 (1999).

Sierra Club Insider. "Sierra Club Brownfields Guidance," editorial. Sierra Club, October 1996.

Simmonds, M. *Cetacean Habitat Loss and Degradation in the Mediterranean Sea*. Report published by ACCOBAMS (Agreement on the Conservation of Cetaceans in the Black Sea Mediterranean Sea and Contiguous Atlantic Area), February 2002.

Sjodin, A., et al. "Retrospective Time Trend Study of PCBs in Human Serum from Various Regions of the United States, 1985–2002." CDC (Centers for Disease Control) online, 2003.

Skakkebaek, N., et al. "Testicular Dysgenesis Syndrome: An Increasingly Common Developmental Disorder with Environmental Aspects." *Human Reproduction* 5 (July 2001).

Sklansky, D. A. "General Electric Co. v. Joiner." David Sklansky (blog), 2003.

Skoloff, B. "County Plans to Vaporize Landfill Trash." Associated Press, September 11, 2006.

Slater, D. "Some People Think Outside the Box: Some Don't Think about Boxes at All." *Sierra*, July/August 2005.

Sloan, L. "Fisheries and Oceans Canada to Begin Recovery Planning for Pacific Killer Whales." Fisheries and Oceans Canada, Pacific Region, news release, February 26, 2004.

Smith, D. "Deep Trouble: A Dying Sea." *Naples* (Florida) *Daily News*, October 3, 2003.

"Sounds Simple Enough." *University of California-San Francisco Magazine* 25, no. 1 (December 1, 2004).

SourceWatch. "Profile of Gina Kolata." www.sourcewatch.org, 2006.

Spitzer, E. "Statement by Attorney General Eliot Spitzer Regarding General Electric's New Request to Further Delay Dredging of the Hudson River." Department of Law, New York State, news release, March 22, 2006.

Squires, S. "Good Fish, Bad Fish." *Washington Post*, August 8, 2006.

Stanley, J. S. "Volume III: Semivolatile Organic Compounds." From *Broad Scan Analysis of the FY82 National Human Adipose Tissue Survey Specimens*, Environmental Protection Agency, 1982.

Stebbins, W. *A Natural History of Amphibians*. Princeton, NJ: Princeton University Press, 1995.

Steinbrecher, R. "Terminator Technology: The Threat to World Food Security" *Ecologist* 28, no. 5 (1998).

Stephen Safe op-eds. Our Stolen Future Web site, www.ourstolenfuture.org/Commentary/Opinion/1997safeserrors.htm.

Stoddard, E. "Pregnancy Test May Lie behind Deadly Frog Fungus." Reuters, February 8, 2006.

Stow, J. "Best Available Scientific Information on the Effects of Deposition of POPs." Northern Contaminants Program, Indian and Northern Affairs, Canada, www.ainc-inac.gc.ca, May 2005.

Sures, B., et al. "Individual and Combined Effects of (PCB 126) on the Humoral Immune Response in European Eel Experimentally Infected with Larvae of Nematoda." *Parasitology* 128 (2004).

Suwol, R. "Mothers' PCBs and Sons' Testicular Cancer." *Rachel's Democracy & Health News* 13 (March 2003).

Suzuki, D. "Need More Research on Environmental Chemical Exposure: A Guest Commentary." Environmental News Network, August 22, 2005.

Tabb, M., et al. "Highly Chlorinated PCBs Inhibit the Human Xenobiotic Response Mediated by the Steroid and Xenobiotic Receptor (SXR)." *Environmental Health Perspectives* 112, no. 2 (February 2004).

———. "New Modes of Action for Endocrine Disrupting Chemicals." *Molecular Endocrinology* 20, no. 3 (2006).

Tangri, N. "Waste Incineration: A Dying Technology." *Gaia Secretariat*, 2003.

Tapper, J. "Bush's EPA Chief Seeks Greener Pastures." Salon.com, May 22, 2003.

Tarkan, L. "For Adults, Allergies Bring a Surprising Twist." *New York Times*, April 18, 2006.

Teuten, E., et al. "Chemical Compounds Found in Whale Blubber Are from Natural Sources, Not Industrial Contamination." Woods Hole Oceanographic Institution, news release. February 18, 2005.

Thillart, G., et al. "Endurance Swimming of the European Eel." *Journal of Fish Biology* 65 (2004).

Thomas, J. "Boycott: Brands and Products to Avoid" *Ecologist* 28, no. 5 (1998).

Thompson, J. "DEW Line Sites Leaking Contaminants." *Nunatsiaq* (Canada) *News.* September 16, 2005.

Thornton, J. *Pandora's Poison.* Cambridge: MIT Press, 2000.

Thrall, L. "Another Danger for Developing Frogs." *Science News.* February 1, 2006.

"Today's News: Toxic Release Inventory." *Sierra Club News*, May 17, 2005.

Tokar, B. "Monsanto: A Checkered History." *Ecologist* 28, no. 5 (1998).

Topping, A. "How Eating Fish during Pregnancy Could Make Baby Brainier." *Guardian Unlimited*, February 16, 2007.

Trubo, R. "Endocrine Disrupting Chemicals Probed." *Journal of the American Medical Association* 203, no. 3 (July 20, 2005).

Tryphonas, H. "Immunotoxicity of PCBs (Aroclors) in Relation to the Great Lakes." *Environmental Health Perspectives* 103, suppl. 9 (December 1995).

———. "Polychlorinated Biphenyl-induced Immunomodulation and Human Health Effects." In *PCBs: Recent Advances in Environmental Toxicology and Health Effects*, eds. Robertson and Hansen.

Turner, D. "Amphibian Ark Planned to Save Frogs." Associated Press, February 15, 2007.

Turque, N., et al. "A Rapid, Physiologic Protocol for Testing Transcriptional Effects of Thyroid Disrupting Agents in Pre-Metamorphic Xenopus Tadpoles." *Environmental Health Perspectives* 113, no. 11 (November 2005).

Tye, L. "Journal Fuels Conflict-of-Interest Debate." *Boston Globe*, January 6, 1998.

"UK Testicular Cancer Incidence Statistics." Cancer Research UK online, 2006.

UNEP (United Nations Environmental Programme). *Inventory of Worldwide PCB Destruction Capacity.* www.unep.org, December 1998.

———. "Survey of Currently Available Non-incineration PCB Destruction Technologies." *UNEP Chemicals* 1 (2000).

———. "Global Clean-Up of Toxic PCBs." News release, June 10, 2004.

University of California News Office. "Researchers Identify How PCBs May Alter in Utero, Neonatal Brain Development." www.universityofcalifornia.edu/news/article/20944, April 13, 2009.

U.S. Court of Appeals. "GE v. EPA Opinion." November 20, 2003.

U.S. Department of Health and Human Services. Agency for Toxic Substances and Disease Registry. "Toxicological Profile for Polychlorinated Biphenyls (PCBs)." November 2000.

U.S. Environmental Protection Agency. "EPA Incinerator Approvals to Speed PCB Disposal." EPA press release, February 10, 1981.

———. Office of Pollution Prevention and Toxics. *Management of Polychlorinated Biphenyls in the United States.* January 30, 1997.

———. "Disposal of Polychlorinated Biphenyls (PCBs): Final Rule." *Federal Register* 63, no. 124 (June 29, 1998).

——— "EPA Releases Hudson River Record of Decision." EPA news release, December 4, 2001.

———. Office of Pollution Prevention and Toxics. "Hazard Assessment." November 4, 2002.

———. "Commercially Permitted PCB Disposal Companies." December 2005.

———. "Public Health Implications of Exposure to Polychlorinated Biphenyls (PCBs)." 2006.

———. "Transformer Databases and Forms." EPA form 7720–12.

———. EPA Region 10. "DOJ, EPA Reach Agreement with General Electric to Conduct Hudson River Dredging." EPA news release, October 6, 2005.

U.S. Food and Drug Administration. "Title 21 CFR—Food and Drugs" The Federal Register, April 1, 2005.

VanGelder, S. "RX for Earth," talk with Dr. Theo Colborn." *Yes!* Magazine, June 30, 1998.

vom Saal, F. Interview by Doug Hamilton. *Frontline*, PBS, June 2, 1998.

vom Saal, F., et al. "Bisphenol A Expert Panel Consensus Statement: Integration of Mechanisms, Effects in Animals and Potential Impact to Human Health at Current Exposure Levels." *Reproductive Toxicology* 131, no. 8 (August-September 2007).

Vreugdenhil, H., et al. "Effects of Perinatal Exposure to PCBs and Dioxins on Play Behavior in Dutch Children at School Age." *Environmental Health Perspectives* 110, no. 10 (September 13, 2002).

Wallberg, P., et al. "Potential Importance of Protozoan Grazing on the Accumulation of Polychlorinated Biphenyls (PCBs) in the Pelagic Food." *Biomedical and Life Sciences* 357, nos. 1–3 (1997).

Walsh, D. B. "The Business Council Opposes Dredging to Remove PCBs in the Upper Hudson River." Business Council of New York State, www.bcnys .org, 2000.

Ward, B. "Is There a Connection between Environmental Toxins and Breast Cancer?" *E* magazine, September 14, 2004.

Waring, T., et al. "U.S. Atlantic Marine Mammal Stock Assessments: 1998." U.S. Department of Commerce, National Oceanic and Atmospheric Administration, National Marine Fisheries Service, Northeast Fisheries Science Center, December 1998.

"Washington State University Study Shows Environmental Toxins Can Cause Inherited Diseases." Washington State University, WSU News Service, Research News, September 14, 2006.

Watson, P. "Toxic Roulette and the Revenge of the Fish." *New Zealand Herald*, August 22, 2006.

Weaver, J. "Breast Cancer Risk Minimized by Breastfeeding." Yale University, news release, June 18, 2001.

Webster, B. "Pollution in Long Island Sound Tied to Defects in Terns." *New York Times*, October 29, 1971.

"A Weight-of-Evidence Assessment of the Human Health Risks of PCBs." General Electric online, 2006.

Weisglas-Kuperus, N., et al. "Immunologic Effects of Background Exposure to Polychlorinated Biphenyls and Dioxins in Dutch Preschool Children." *Environmental Health Perspectives* 108, no. 12 (December 2000).

Weisskopf, C. "Here's Another Fine Mess You've Gotten Us Into." *Agrichemical and Environmental News Index*, May 1998.

Welch, C. "Toxins Permeate All Levels of Marine Life, Report Says." *Seattle Times*, December 12, 2001.

———. "Endangered Listing for Elusive Orcas? Scientists Seek Clues to Decline." *Seattle Times*, December 19, 2003.

Weldon, C., et al. "Origin of the Amphibian Chytrid Fungus." *Emerging Infectious Diseases* 10, no. 12 (December 2004).

Whelan, E. M. *Toxic Terror.* Amherst, NY: Prometheus Books, 1993.

————. Op-ed. Our Stolen Future.org, December 2000.

Whitaker, J. C. Speech to the Nature Conservancy, Boulder, Colorado, November 19, 2003.

Wienhold, R. "Focus: Epigenetics." *Environmental Health Perspectives* 114, no. 3 (March 2006).

Williams, F. "Toxic Breast Milk." *New York Times.* January 9, 2005.

Williams, J. C. "GE Wins Challenge against USEPA Superfund Order." May It Please the Court: A Weblog of Legal News & Observations, March 5, 2004.

Wilson, E. O. *The Future of Life.* New York: Borzoi Books, 2002.

Wilson, M. "Dousing the Flames: Communities Unite Globally to Lock Out the Incinerator Industry." *Multinational Monitor* 25, no. 1/2 (2004).

Wisconsin Electricity Coordinating Council. "Global Modeling of PCBs." *WECC Report,* January 1999.

Wolf, V. "Global Chemical Contamination Threatens Child Development: A Guest Commentary." Environmental News Network, August 19, 2005.

Wong, C. S. "Organochlorine Pesticides and PCBs in Stream Sediment and Aquatic Biota." U.S. Geological Survey, 2002.

Wong, K. "Amphibians Are Threatened Worldwide and Other Stories." Environmental News Network, November 5, 2004.

World Wildlife Fund. "Contamination: The Results of the WWF's Biomonitoring Survey." www.worldwildlife.org, 2003.

Wright, S. "Selling Food, Health, Hope." RESIST (Resistance and Solidarity against Agrochemical TNCs), www.pressurepoint.org. June 4, 2003.

Yang, D., et al. "Developmental Exposure to Polychlorinated Biphenyls Interferes with Experience-Dependent Dendritic Plasticity and Ryanodine Receptor Expression in Weanling Rats." *Environmental Health Perspectives* 117, no. 3 (March 2009).

Yoon, C. "In Nurses' Lives, a Treasure Trove of Health Data." *New York Times,* September 15, 1998.

Yoshimura, T. "Yusho in Japan." *Industrial Health* 41 (2003).

Zafar, A. "Conserving Our Coastal Environment." United Nations University, 2002.

Zhang, Y., et al. "Serum Polychlorinated Biphenyls, Cytochrome P-450 1A1 Polymorphisms and Risk of Breast Cancer in Connecticut Women." *American Journal of Epidemiology* 160 (2004).

Zheng, T., et al. "Breast Cancer Risk Associated with Congeners of PCBs." *American Journal of Epidemiology* 152, no. 1 (2000).

Index